Lecture Notes in Networks and Systems 1009

Series Editor

Janusz Kacprzyk ⓘ, *Systems Research Institute, Polish Academy of Sciences, Warsaw, Poland*

Advisory Editors

Fernando Gomide, *Department of Computer Engineering and Automation—DCA, School of Electrical and Computer Engineering—FEEC, University of Campinas—UNICAMP, São Paulo, Brazil*

Okyay Kaynak, *Department of Electrical and Electronic Engineering, Bogazici University, Istanbul, Türkiye*

Derong Liu, *Department of Electrical and Computer Engineering, University of Illinois at Chicago, Chicago, USA*

  Institute of Automation, Chinese Academy of Sciences, Beijing, USA

Witold Pedrycz, *Department of Electrical and Computer Engineering, University of Alberta, Alberta, Canada*

  Systems Research Institute, Polish Academy of Sciences, Warsaw, Canada

Marios M. Polycarpou, *Department of Electrical and Computer Engineering, KIOS Research Center for Intelligent Systems and Networks, University of Cyprus, Nicosia, Cyprus*

Imre J. Rudas, *Óbuda University, Budapest, Hungary*

Jun Wang, *Department of Computer Science, City University of Hong Kong, Kowloon, Hong Kong*

The series "Lecture Notes in Networks and Systems" publishes the latest developments in Networks and Systems—quickly, informally and with high quality. Original research reported in proceedings and post-proceedings represents the core of LNNS.

Volumes published in LNNS embrace all aspects and subfields of, as well as new challenges in, Networks and Systems.

The series contains proceedings and edited volumes in systems and networks, spanning the areas of Cyber-Physical Systems, Autonomous Systems, Sensor Networks, Control Systems, Energy Systems, Automotive Systems, Biological Systems, Vehicular Networking and Connected Vehicles, Aerospace Systems, Automation, Manufacturing, Smart Grids, Nonlinear Systems, Power Systems, Robotics, Social Systems, Economic Systems and other. Of particular value to both the contributors and the readership are the short publication timeframe and the world-wide distribution and exposure which enable both a wide and rapid dissemination of research output.

The series covers the theory, applications, and perspectives on the state of the art and future developments relevant to systems and networks, decision making, control, complex processes and related areas, as embedded in the fields of interdisciplinary and applied sciences, engineering, computer science, physics, economics, social, and life sciences, as well as the paradigms and methodologies behind them.

Indexed by SCOPUS, INSPEC, WTI Frankfurt eG, zbMATH, SCImago.

All books published in the series are submitted for consideration in Web of Science.

For proposals from Asia please contact Aninda Bose (aninda.bose@springer.com).

Herwig Unger · Marcel Schaible
Editors

Advances in Real-Time and Autonomous Systems

Proceedings of the 15th International Conference on Autonomous Systems

 Springer

Editors
Herwig Unger
Lehrgebiet Kommunikationsnetze
FernUniversität in Hagen
Hagen, Germany

Marcel Schaible
Lehrgebiet Kommunikationsnetze
FernUniversität in Hagen
Hagen, Nordrhein-Westfalen, Germany

ISSN 2367-3370   ISSN 2367-3389 (electronic)
Lecture Notes in Networks and Systems
ISBN 978-3-031-61417-0   ISBN 978-3-031-61418-7 (eBook)
https://doi.org/10.1007/978-3-031-61418-7

This Springer imprint is published by the registered company Springer Nature Switzerland AG
The registered company address is: Gewerbestrasse 11, 6330 Cham, Switzerland

If disposing of this product, please recycle the paper.

Preface

It is with great pleasure that we present the second edition of the "Real-time and Autonomous Systems" volume, published by Springer in English. This collection showcases the advancements made in this field by European and Thai scientists. The conference on "Autonomous Systems" has now reached its 15th iteration, with a brief interruption due to the COVID-19 pandemic.

Originating in 2008 as a modest workshop for PhD students, our conference has evolved into a cherished tradition. Since 2010, colleagues from diverse research domains have contributed articles on their ongoing work, unsolved scientific challenges, and research outcomes to our proceedings. The event was initially titled "Distributed Systems and Networks", but in subsequent years, it was aptly renamed "Autonomous Systems" to encompass the broad spectrum of contributions. This term encapsulates self-contained and self-controlled entities that operate without external oversight in various scientific disciplines.

In 2019, we made the significant decision to rebrand the publication from "proceedings" to "almanac", reflecting its evolving nature. Authors are now encouraged to contribute without the obligation to attend or present at the conference. Maintaining our commitment to openness, the almanac remains free from reviews or censorship, fostering a platform for unconventional ideas and perspectives. Distributed among conference participants, it serves as a catalyst for uninhibited discussions, emphasizing the importance of intellectual diversity in scientific discourse.

In 2023, we ventured into a collaborative publication by combining the proceedings of two conferences: the "Real-Time Systems" in Boppard and the "Autonomous Systems" in Majorca Island. Regrettably, we couldn't sustain this collaboration due to the cancelation of the "Real-time Systems" meeting in 2023 owing to insufficient submissions. Despite challenges, the "Autonomous Systems" conference in 2023 flourished with over 40 participants, reminiscent of pre-pandemic times. Notable keynote speakers, including Wookey Lee (Korea), Wolfram Schiffmann (Germany), and Phayung Meesad (Thailand), enriched our program with insights into team synergy, emergency landing systems, and stock analysis using deep reinforcement learning. Stephan Pareigis (Germany) provided a comprehensive tutorial on Reinforcement Learning, while Kyandoghere Kyamakya showcased groundbreaking results in "Intelligent Traffic Systems". Additionally, for the first time a tutorial on the real-time, safety-related programming language PEARL by Wolfgang Halang is published in English in this volume.

Innovating our presentation format, we introduced open discussion sessions for major topics, fostering interactive exchanges among participants. This format, characterized by brief talks followed by small-group discussions involving specialists and non-specialists, received positive feedback, indicating a promising future for interactive conference sessions.

The editors hope this publication inspires and informs readers, sparking new ideas and expanding knowledge horizons. We extend a warm invitation to join us at the upcoming conference in Cala Millor, Spain, from October 22–27 2024, either as a participant or a contributor. Further details are available at https://www.confautsys.org.

We eagerly anticipate your engagement and contribution to our vibrant scientific community.

January 2024

Herwig Unger
Marcel Schaible

Organization

Program Commitee

R. Baran, Hamburg
J. Bartels, Krefeld
M. Baunach, Graz
B. Beenen, Lüneburg
J. Benra, Wilhelmshaven
V. Cseke, Wedemark
R. Gumzej, Maribor
W.A. Halang, Hagen
H.H. Heitmann, Hamburg
P. Holleczek, Erlangen
M.M. Kubek, Hagen
Z. Li, Hagen
R. Müller, Furtwangen
S. Pareigis, Hamburg
M. Pellkofer, Landshut
M. Schaible, München
G. Schiedermeier, Landshut
U. Schneider, Mittweida
D. Tutsch, Wuppertal
H. Unger, Hagen (Vorsitz)
C. Yuan, Köln
D. Zöbel, Koblenz

Internet Address of the Real-Time Systems Committee: www.real-time.de
CR Subject Classification (2001): C3, D.4.7

Contents

Algorithmic Foundations
of Reinforcement Learning

Stephan Pareigis[✉]

Department of Informatics, Hamburg University of Applied Sciences, Berliner Tor 7,
20099 Hamburg, Germany
`stephan.pareigis@haw-hamburg.de`

Abstract. A comprehensive algorithmic introduction to reinforcement
learning is given, laying the foundational concepts and methodologies.
Fundamentals of Markov Decision Processes (MDPs) and dynamic programming are covered, describing the principles and techniques for
addressing model-based problems within MDP frameworks. The most
significant model-free reinforcement learning algorithms, including Q-
learning and actor-critic methods are explained in detail. A comprehensive overview of each algorithm's mechanisms is provided, forming a
robust algorithmic and mathematical understanding of current practices
in reinforcement learning.

Keywords: reinforcement learning · MDP · markov-decision process ·
dynamic programming · deep reinforcement learning · SARSA ·
Q-learning · DQN · REINFORCE · A2C · PPO · DDPG · SAC ·
policy gradient methods · exploration vs exploitation · sparse rewards ·
robotics · offline reinforcement learning

1 Introduction

The article provides a comprehensive introduction and overview over the algorithmic foundations of Reinforcement Learning (RL).

Reinforcement Learning forms an important part of artificial intelligence,
characterized by learning optimal decision-making through interactions with
dynamic environments. The field is characterized by numerous technological
applications, including autonomous systems, strategic game playing, and complex decision-making processes.

Figure 1 illustrates the algorithmic and theoretical roots of reinforcement
learning, its categorization within the area of artificial intelligence, and important fields of application. RL can be categorized as a distinct type of machine
learning, next to supervised learning and unsupervised learning. Its algorithmic
roots lie within optimal control theory and dynamic programming. Applications
include problems in which sequential decisions have to be made in order to optimize a given objective function.

© The Author(s), under exclusive license to Springer Nature Switzerland AG 2024
H. Unger and M. Schaible (Eds.): AUTSYS 2023, LNNS 1009, pp. 1–27, 2024.
https://doi.org/10.1007/978-3-031-61418-7_1

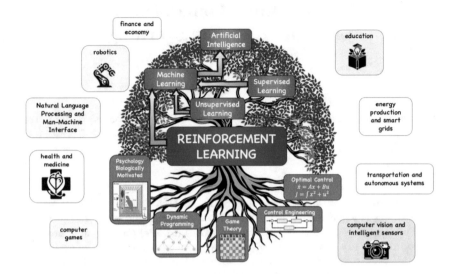

Fig. 1. The Roots of Reinforcement Learning: A visual map illustrating the interdisciplinary nature of Reinforcement Learning (RL). RL is considered a subfield of Artificial Intelligence. Its relationship with Machine Learning, Supervised and Unsupervised Learning, and connections to Dynamic Programming, Game Theory, and Control Engineering. The diagram further branches out to show RL's diverse applications in sectors such as healthcare, finance, robotics, energy, education, computer vision, and more, underlining its profound impact across industries.

A comprehensive book on reinforcement learning is R. Sutton's and A. Barto's book Reinforcement Learning: An Introduction [1]. Other books which cover the basics of reinforcement learning including practical examples and modern research areas and applications are M. Lapan's book on Deep Reinforcement Learning [2] and the workshop book from Palmas et al. [3]. [4] gives theoretical background on modern Deep RL methods like PPO and A2C.

Section 2 gives an introduction to the theoretical concepts of reinforcement learning. Section 2.1 covers the key concept of a Markov Decision Process (MDP). The problem setting is to find an optimal strategy for the MDP in form of a policy which optimizes the total reward. If full knowledge of the MDP is given, dynamic programming principles can be applied to obtain an optimal strategy. Section 2.2 explains how the Bellman Equation is used to iteratively approximate a solution.

If no model of the MDP is given, then exploration methods must be used to learn from the interactions with the MDP as covered in Sect. 3. The methods are based upon dynamic programming principles as described previously, and they can be divided into on-policy and off-policy methods.

When observation spaces are based on image inputs from cameras or are otherwise too large to handle, then artificial neural networks are used to approximate an optimal strategy. This area is referred to as deep reinforcement learning.

Obtaining a solution for the optimization problem can either be done based on the approximation of an action-value function as described in Sect. 3.2 or the approximation of a policy as explained in Sect. 3.3. A combination of value and policy based methods are actor-critic methods as described in Sect. 3.4.

If the action space is continuous, then specially designed methods are used as carried out in Sect. 3.5.

2   Reinforcement Learning Fundamentals

The theoretical foundation of reinforcement learning is based upon Markov Decision Processes (MDPs), which provide a mathematical framework for modeling decision-making in stochastic environments. The central objective is to identify an optimal decision-making strategy that maximizes a specified objective function. The chapter covers the definition of an MDP, derives a policy-induced trajectory distribution, introduces value functions, and the Bellman Equation. The chapter closes with dynamic programming principles which are used to obtain a value function given full knowledge of the underlying MDP.

2.1   Markov Decision Processes (MDPs)

The following basic definitions from stochastic processes are frequently used in the formulation of MDPs and Reinforcement Learning.

Definition. The *conditional probability* of an event A given an event B is defined as $P(A|B) := \frac{P(A \cap B)}{P(B)}$.

If events A and B are independent, then $P(A|B) = P(A)$. In this case, the probability that both events A and B occur can be calculated as

$$P(A \cap B) = P(A) \cdot P(B). \tag{1}$$

The product rule for independent events is essential for calculations in MDPs.

Definition. If X is a random variable and p is its probability distribution, then the *expected value* of X is defined as

$$\mathbb{E}[X] := \int x \cdot p(X = x)\, dx \quad \text{or} \quad \mathbb{E}[X] := \sum_x x \cdot p(x) \tag{2}$$

depending on whether X is continuous or discrete.

Elements of an MDP. An MDP is composed of the following components:

- **States ($\mathcal{S}$):** A state represents a specific situation in the environment, the physical world, or system to be controlled. A state space may be high dimensional like the raw input from a camera. It may also be low dimensional like a 2D grid or a feature space. An MDP models a system that transitions between different states.

- **Observations** (O): Ideally all states can be fully observed. In this case $S = O$. If not all states of an MDP can be fully observed, then it is called a *partially observable MDP* or POMDP. The *observation space* is used in RL to describe the set of all possible observable input (sensor input, feature information, odometric information, etc.).
- **Actions** (A): Actions are the decisions or moves that an agent can take in a given state. The set of all possible actions in a state is denoted as $A(s)$.
- **Transitions** $\mathcal{P}$: The transition probability function

$$P : \mathcal{S} \times A \times \mathcal{S} \to [0, 1], \qquad (s, a, s') \mapsto P(s'|s, a) \in [0, 1] \tag{3}$$

describes the probability of transitioning from one state to another given a particular action. It is often represented as $P(s'|s, a)$, denoting the probability of transitioning to state s' from state s by taking action a.

The Markov property of an MDP states that the transition probability to a state s_{t+1} is only dependent on the predecessor state s_t and action a_t, not on past states, i.e. a MDP is memory-less

$$p(s_{t+1}|s_t, a_t) = p(s_{t+1}|a_t, s_t, a_{t-1}, s_{t-1}, \cdots).$$

- **Rewards** R: Each state-action-state triple receives a numerical reward

$$R : \mathcal{S} \times A \times \mathcal{S} \to \mathbb{R}.$$

The immediate reward is denoted as $R(s'|s, a)$. The expected immediate reward in a state s with action a can be expressed as

$$\mathbb{E}[R(s, a)] = \sum_{s' \in \mathcal{S}} P(s'|s, a) \cdot R(s'|s, a) \tag{4}$$

- **Policy** (π): A policy is a strategy that the agent follows to choose actions in each state. It can be deterministic

$$\pi : S \longrightarrow A, \quad \pi(s) = a \in A \tag{5}$$

or stochastic

$$\pi : A \times S \longrightarrow [0, 1], \quad \pi(a|s) = p \in [0, 1]. \tag{6}$$

When a policy π is given, the expected immediate reward in a state s can be written as

$$\mathbb{E}[R(s)] = \sum_{s' \in S, a \in A} \pi(a|s) \cdot P(s'|s, a) \cdot R(s'|s, a) \tag{7}$$

- **Realizing an MDP:** A sequence of states $\tau = (s_0, a_0, r_0, s_1, a_1, r_1, \cdots, s_T)$ is called a realization of an MDP for some starting state s_0, when a_t is chosen according to π, i.e.

$$a_t = \pi(s_t) \text{ or choosen from the distribution } a_t \sim \pi(.|s_t)) \tag{8}$$

and $r_t = R(s_{t+1}|s_t, a_t)$. τ is also called trajectory or path. T can be a finite or infinite time horizon.

- **Total Return:** The cumulated reward or total return G for some realization τ is defined as

$$G := \sum_{t=0}^{T-1} \gamma^t R_t, \quad R_t = R(s_{t+1}|s_t, a_t) \text{ along } \tau \tag{9}$$

for some discount factor $\gamma \in \ ]0,1]$. The goal in RL is to find a policy $\pi^\star$ which maximizes the expected total reward $\mathbb{E}_{\tau \sim \pi}[G]$. $\tau \sim \pi$ denotes a realization of a trajectory given a policy π.

The discount factor γ may be set to 1 for MDPs which terminate after a finite time. For infinite horizon MDPs a discount factor $\gamma < 1$ is necessary to ensure existence of G.

Policy Induced Trajectory Distribution $\tau \sim \pi$. For a given policy π every concrete trajectory $\tau = (s_0, a_0, r_0, s_1, a_1, r_1, \cdots, s_T)$ has a certain probability. The probability distribution for trajectories for a fixed policy π will be derived in this section.

According to (3) the probability for ending up in state s_1 when starting in state s_0 with a given action a_0 is $p(s_1|s_0, a_0)$. If a_0 is chosen according to the given policy $a_0 \sim \pi(.|s_0)$, then the probability for the sequence (s_0, a_0, s_1) is

$$P((s_0, a_0, s_1)) = \pi(a_0|s_0) \cdot P(s_1|s_0, a_0) \tag{10}$$

Policy Induced Trajectory Distribution. Using the memory-lessness of an MDP together with the product rule for independent probabilities (1), the probability for a specific trajectory τ_{s_0} starting in state s_0 can be calculated as the product of the probabilities in (10)

$$P_\pi(\tau) = P_\pi(s_0, a_0, \cdots, s_T) = \prod_{t=0}^{T-1} \pi(a_t|s_t) \cdot P(s_{t+1}|s_t, a_t) \tag{11}$$

Equation (11) is called the *policy induced trajectory distribution* for $\tau \sim \pi$.

This probability can be used to calculate the expected value of the total return $\mathbb{E}_{\tau \sim \pi}[G]$ over all trajectories τ starting in state s_0.

Equation (12A) uses the definition of the expected value. (12B) shows that $\mathbb{E}_{\tau \sim \pi}[G]$ may be expressed as expected values of local rewards, using the *occupancy measure d^π*.

Using (9) and (11) the expected total return for G_π can be calculated as

$$\mathbb{E}_{\tau \sim \pi}[G_\pi] \overset{A}{=} \sum_\tau P_\pi(\tau) \cdot G_\pi(\tau) \overset{B}{=} \sum_s d^\pi(s) \cdot \mathbb{E}[R(s)] \tag{12}$$

Occupancy Measure. The value d^π as introduced in Eq. (12) is called the *discounted occupancy measure*.

$$d^\pi(s) = \mathbb{E}_{\tau \sim \pi} \left[\sum_{t=0}^{T-1} \gamma^t \cdot \mathbb{I}(s_t = s) \right]$$

where $\mathbb{I}(s_t = s)$ is 1 if the $s_t = s$. The value d^π expresses the discounted frequency of visiting state s given a policy π. Figure 2 explains the property (12B) in a simple example. The expected total return in state A can hence be expressed in two different ways.

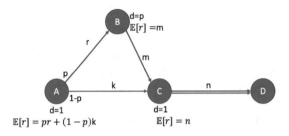

Fig. 2. The example illustrates Eq. (12) and shows a Markov Process which allows a red path $\tau_{\text{red}} = (A, B, C, D)$ and a blue path $\tau_{\text{blue}} = (A, C, D)$. The rewards are k, r, m, n as shown on the edges of the graph. The probability for τ_{red} is $P(\tau_{\text{red}}) = p \cdot 1 \cdot 1$. The probability for τ_{blue} is $P(\tau_{\text{blue}}) = (1-p) \cdot 1$. The expected total return for A is therefore $\mathbb{E}[G] = p \cdot (r + m + n) + (1 - p) \cdot (k + n)$. Using the occupancy measure and expected one-step rewards in each state as shown in the image, alternatively the expected return can be calculated as $\mathbb{E}[G] = \sum_{s \in S} d(s) \cdot \mathbb{E}[r(s)] = 1 \cdot (pr + (1 - p)k) + p \cdot m + 1 \cdot n$.

The following sections will deal with the expected total return, which will be called value function V, and methods for calculating it.

Value Functions. The objective is to find an optimal policy that maximizes the expected cumulative reward or total return. If in every state $s \in S$ a prediction of the expected total return G_s were known, then an optimal action a could be chosen which leads to a subsequent state in which the prediction of the expected total return is maximized. Such a prediction is called a state value function V.

State Value Function (V): The *state value function* $V(s)$ for a policy π in an MDP is defined as the expected total return starting from state s and following policy π thereafter. Mathematically, it can be expressed as:

$$V^\pi(s) := \mathbb{E}_{\tau \sim \pi}\left[\sum_{t=0}^{T-1} \gamma^k R_t \mid S_0 = s\right] \tag{13}$$

In order to be able to calculate a greedy action or policy based on the state value function V^π, the transition properties of the MDP must be known. A greedy policy $\pi^{*,\pi}$ with respect to some value function V^π (based on a fixed policy π) is then calculated as

$$\pi^{*,\pi}(s) = \arg\max_{a \in A}\left[\sum_{s' \in S} P(s'|s, a)[R(s, a, s') + \gamma V^\pi(s')]\right] \tag{14}$$

In each state s the action $a^* = \pi^{*,\pi}(s)$ is chosen, which maximizes the expected total return when taking one step with a^* and continuing with policy π thereafter. Note that $\pi^{*,\pi}$ is generally not an optimal policy. It is a greedy policy based on the value function V^{π} of given policy π.

If the transition properties of the MDP are not known, then the action value function Q^{π} can serve as a prediction for the expected total return.

Action Value Function (Q): The *action-value function* $Q(s, a)$, for a policy π, in an MDP is defined as the expected return starting from state s, taking an action a, and thereafter following policy π. It can be formally written as:

$$Q^{\pi}(s, a) = \mathbb{E}_{\tau \sim \pi} \left[\sum_{t=0}^{T-1} \gamma^k R_t \mid S_t = s, A_t = a \right] \tag{15}$$

The difference to the above is, that Q also depends on an action a. This way an optimal action can be chosen by taking the action a^* which maximizes the Q-value in the current state s:

$$a^* = \arg \max_{a \in A} Q(s, a) \tag{16}$$

Comparing (14) and (16) shows that in the latter case no transition information of the MDP is necessary in order to choose a greedy action in every state.

Bellman Equations. The Bellman Equations form an essential component of the algorithmic structure of dynamic programming and reinforcement learning. They express that the value of a state can be expressed as an immediate reward plus the value in the next state.

The **Bellman Expectation Equation** for the **state value function V** given a policy π is given by

$$V^{\pi}(s) = \sum_{a \in A} \pi(a|s) \cdot \sum_{s' \in S} P(s'|s, a)[R(s, a, s') + \gamma V^{\pi}(s')] \tag{17}$$

The **Bellman Expectation Equation** for the **action value function Q** given a policy π is given by

$$Q^{\pi}(s, a) = \sum_{s' \in S} P(s'|s, a) \cdot \left[R(s, a, s') + \gamma \sum_{a' \in A} \pi(a'|s')Q^{\pi}(s', a') \right] \tag{18}$$

The **Bellman Optimality Equation** for the **state value function** is:

$$V^*(s) = \max_{a \in A} \sum_{s' \in S} P(s'|s, a) \cdot [R(s, a, s') + \gamma V^*(s')] \tag{19}$$

The **Bellman Optimality Equation** for the **action value function Q** is given by:

$$Q^*(s, a) = \sum_{s' \in S} P(s'|s, a) \cdot [R(s, a, s') + \gamma \max_{a' \in A} Q^*(s', a')] \tag{20}$$

The Bellman Equations form a basis for the following algorithms.

2.2   Solving an MDP: Dynamic Programming

If the transition information of the MDP is known, then Eq. (17) can be applied iteratively to calculate an optimal policy π^*. The methods are explained in the following section.

Policy Iteration and Value Iteration. In **Policy Iteration** an optimal policy is directly calculated using the following two steps:

1. **Policy Evaluation:** Compute V^π for the current policy, iteratively applying the Bellman Expectation Equation (17).
2. **Policy Improvement:** Update the policy π to a greedy policy $\pi^{*,\pi}$ by selecting in every state the action which maximizes the expression in the Eq. (14).
3. Repeat until convergence.

Figure 3a shows the progression of the algorithm for a simple grid example. The heat maps in the top row show the value function, the bottom row shows the greedy policy with respect to the current value function.

In **Value Iteration** the optimal value function V^* is calculated directly using iterative application of the Bellman Optimality Equation (19). It can be said that policy evaluation and policy improvement are integrated into a single update, iterating until convergence as shown in Fig. 3b.

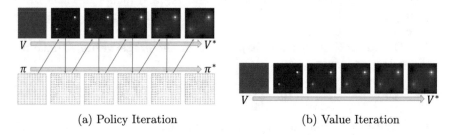

(a) Policy Iteration   (b) Value Iteration

Fig. 3. In *policy iteration* (left) a policy is evaluated and then improved acting greedily. Often it is not necessary to calculate the value function precisely as the policy converges fast to an optimal policy. In *value iteration* (right) the optimal value function is directly approximated using the Bellman Optimality Equation (19).

Dynamic Programming Iteration Methods: There exist various methods for the sequence in which the states are be updated when applying policy evaluation or value iteration. The methods are depicted in Fig. 4.

Synchronous Backups: In each iteration, the algorithm computes the new value for every state based on the old values (from the previous iteration) and updates them all at once at the end of the iteration. This method ensures that the updates in one iteration are independent of each other, providing stability in the convergence process.

In-Place Dynamic Programming: The algorithm updates the value of a state as soon as the new value is computed, using these updated values for subsequent state value computations within the same iteration. It can lead to faster convergence as it uses the most updated information available.

Prioritized Sweeping: The algorithm prioritizes updates for states where the value change is expected to be largest. This way, it uses computational resources more effectively.

Real-Time Dynamic Programming (RTDP): RTDP involves executing trials (realizations) updating the value of states that are encountered in these trials. It is particularly useful in large or continuous state spaces and where the full model of the environment is not available.

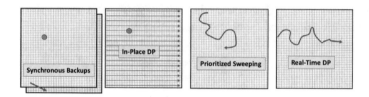

Fig. 4. Different methods for iterating through the state space: Synchronous Backups, In-Place Dynamic Programming, Prioritized Sweeping, RTDP. All the methods require that the transition information of the MDP is given.

2.3   Model-Free Prediction

The previous methods were based on a given model of the MDP. The iterative application of the Bellman Equation required knowledge of the transition properties of the MDP. In the model-free case, value functions have to be learned through observation of the system. These methods can be categorized into *learning from complete episodes* and *learning in temporal differences*. Figure 5 figuratively shows the difference between the methods dynamic programming (DP), Monte Carlo Methods (MC), and temporal difference methods (TD).

Learning from Complete Episodes: Monte Carlo Method (MC) estimate the value function based on complete episodes of experience. The value of a state is computed as the average of the returns following that state over many episodes. An episode must terminate for MC methods to update the value function, making them suitable for episodic tasks.

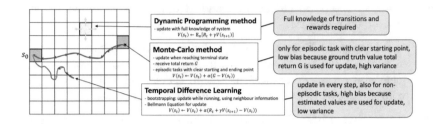

Fig. 5. Overview of update methods: The diagram illustrates three strategies. Dynamic Programming (DP) depicted by the yellow arrows requires full knowledge of state transitions and rewards. Monte Carlo methods learn from complete episodes and update values based on an empirical mean of total returns. Temporal Difference Learning, also known as bootstrapping, updates values incrementally using neighbor information, applicable to both episodic and non-episodic tasks.

Update Rule. The MC updates the value function in every state using an empirical mean calculated from the total returns

$$V_{n+1} \leftarrow \frac{1}{n} \cdot \sum_{i=1}^{n} G_i = V_n + \frac{1}{n}(G_n - V_n) \qquad (21)$$

Equation (21) shows that the update method can be implemented using only the previous value of the value function approximation V_n, the number of visits n and the current total return G_n (Table 1).

Table 1. Comparison of Monte Carlo and Temporal Difference Learning

Aspect	Monte Carlo (MC)	Temporal Difference (TD)
Definition	Empirical average returns from complete episodes	Empirical estimation of the value function based on the rewards and estimated values of subsequent states.
Evaluation	$V_{n+1} \leftarrow \alpha \cdot (G_n - V_n)$	Bellman Expectation Equation $V(s_t) \leftarrow V(s_t) + \alpha[R_{t+1} + \gamma V(s_{t+1}) - V(s_t)]$
Bias	Unbiased estimates of the value function, as they rely on actual returns observed in complete episodes	Biased estimates due to bootstrapping (using estimates to make further estimates), but can lead to faster convergence.
Variance	High variance, as estimates are directly influenced by the complete returns of individual episodes	Lower variance compared to MC, as updates are more incremental and smoothed over multiple steps

Bootstrapping: Temporal Difference Learning. Temporal Difference (TD) learning combines the Monte Carlo idea of updating empirically from observed values and Dynamic Programming (DP) ideas using adjacent information for an update. The fact that TD learning uses estimates based on other learned estimates is known as *bootstrapping*.

Update Rule TD(0). The update rule for the value function $V(s)$ is given by:

$$V(s_t) \leftarrow V(s_t) + \alpha \left[R_{t+1} + \gamma V(s_{t+1}) - V(s_t) \right]$$

TD(λ) Learning. TD(λ) is an extension of basic TD(0) in which more information from further away along the trajectory is involved in the update process. To implement this idea the concept of *eligibility traces* is introduced. These are temporary records that assign credit to states based on how recently and frequently they have been visited. Eligibility traces decay over time, controlled by the parameter λ, which ranges from 0 to 1. Figure 13c shows the decay rate for eligibility traces of different lengths.

3 Reinforcement Learning Algorithms

This section covers RL algorithms. In contrast to the assumptions in the previous chapter it is now assumed that the agent has no model or knowledge of the underlying MDP.

3.1 Model-Free Reinforcement Learning

Unlike model-based approaches, model-free methods such as SARSA and Q-Learning enable agents to learn optimal policies directly from interactions with the environment. Table 2 shows an overview over the methods.

ε-Greedy Exploration. Model-free RL methods need a way to explore the state space. This can be done using *ε-greedy exploration*. An ε-greedy policy chooses a greedy action and with probability ε it chooses a random action:

$$\pi(a|s) = \begin{cases} \arg\max_{a \in \mathcal{A}(s)} Q(s,a) & \text{with probability } 1 - \epsilon, \\ \text{a random action from } A(s) & \text{with probability } \epsilon. \end{cases} \tag{22}$$

Another exploration method which is generally superior to ε-greed exploration is explained in Fig. 8.

SARSA: On-Policy Learning. SARSA is an on-policy RL algorithm that updates the Q-values based on the current policy. The Q-value update rule is based on the Bellman Equation (18) and is given by:

$$Q(s,a) \leftarrow Q(s,a) + \alpha \left[R + \gamma Q(s',a') - Q(s,a) \right] \tag{23}$$

Table 2. Categorization of Model-Free RL Methods. Typical algorithms are mentioned in the Subcategory: SARSA, Q-learning, and REINFORCE. Actor-Critic methods use a combination of both and are described in Sect. 3.4.

Category	Subcategory	Definition
Value-Based	**On-Policy** SARSA	Estimation of the value function, using data from the policy currently being improved.
	Off-Policy Q-learning	Estimation of the value function, using data from a different policy than the one being optimized.
Policy-Based	**On-Policy** REINFORCE	Directly learns the policy function using data generated by the current policy.
	Off-Policy	Directly learns a policy function from experiences gathered by a different policy, allowing learning from past behaviors.
Actor-Critic	**On-Policy** PPO	Learn from experiences under current policy keeping policy and value function aligned.
	Off-Policy DDPG, SAC	Uses behaviour policy to learn, has to correct the differences in probability distribution using importance sampling

In its general form SARSA uses ε-greedy exploration with decreasing ε during the training process.

SARSA is on-policy because it updates its Q-values based on the actions chosen by its current policy (ε-greedy). This makes SARSA more sensitive to the exploration-exploitation trade-off.

Q-Learning: Off-Policy Learning. The Q-learning update rule is based on the Bellman Equation (20) and is given by

$$Q(s,a) \leftarrow Q(s,a) + \alpha \left[R + \gamma \max_{a'} Q(s',a') - Q(s,a) \right] \qquad (24)$$

In its general form Q-learning also uses ε-greedy exploration with decreasing ε during the training process. However, in the update rule (24) the update is made with a greedy action, as opposed to a ε-greedy action in SARSA. The update rule therefore updates in a greedy way whereas the exploration is done with random actions with probability ε. Q-learning is therefore an off-policy RL algorithm, whereas SARSA learns the action-value function for an ε-greedy policy.

Figure 6 illustrates the differences between on-policy and off-policy learning. On-policy methods converge slower and are more conservative in the policy they evaluate. This is beneficial in environments where taking certain actions can lead to disastrous outcomes. Historic experiences are useless in the training process.

Off-policy methods converge faster because they can learn from experiences generated by a different policy. Replay buffers with historic data may be used effectively.

The following Table 3 compares the two methods on-policy and off-policy.

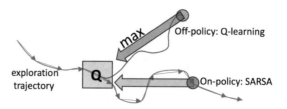

Fig. 6. Comparison between off-policy Q-learning and on-policy SARSA. It illustrates an exploration trajectory taken by an agent (blue), where the Q-learning approach selects the action that maximizes the Q-value (hence "off-policy" as it may choose an action not from the current policy), whereas SARSA selects the next action based on the current policy (hence "on-policy") and updates its Q-values based on the action actually taken.

Table 3. Comparison of Off-Policy and On-Policy Learning

	Off-Policy	On-Policy
Properties	Learns from experiences generated by a different policy than the one being improved	Learns from the experiences generated by the current policy.
Advantages	- Can use data from previous policies - Efficient use of experience (replay buffer) - Suitable for Q-learning and DDPG - Can learn optimal policy while following an exploratory or safe policy	- Updated with the most recent data - Tight coupling between the policy being evaluated and the policy generating the data - Suitable for algorithms like A3C and PPO
Disadvantages	- Can be less stable due to the variance from different behavior policies - Requires careful handling of the importance sampling	- Less efficient use of data as it cannot learn from old experiences - Can be slower to converge as it continually updates from the same policy's experiences

3.2   Deep Q-Learning

Q-learning has initially been used for problems with small state spaces for which grids or table based methods can be used to approximate the action value function. For large state spaces, artificial neural networks can be used to approximate the action value function.

Artificial neural networks (ANN) are designed to map input values to output values. In reinforcement learning, states (from observations) have to be mapped to values of an action value function. However, during the training process and the exploration phase the output values vary as new experiences are made. This leads to stability problems when training ANNs. Deep Q-learning [5] is an algorithm which accounts for the mentioned stability problems.

Figure 7 shows the topology of the ANN. The following components assist in making the training process stable:

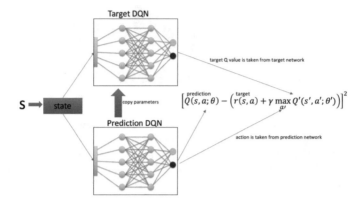

Fig. 7. The diagram illustrates the architecture of a Deep Q-Network (DQN), showing the interaction between two neural networks: the prediction network and the target network. The input state S is fed into both networks. The prediction network is responsible for estimating the Q-value $Q(S, a; \theta)$ for the current state and action. The target network, which has its parameters periodically updated from the prediction network, provides a stable target Q-value for the next state $Q(S', a'; \theta')$. The learning process involves minimizing the loss, which is the squared difference between the prediction network's Q-value and the target Q-value (adjusted by the reward $r(S, a)$ and the discounted maximum Q-value of the next state from the target network). The action to be taken is determined by the prediction network, ensuring a continuous learning and updating cycle for the DQN.

- **Experience Replay Buffer.** Past experiences of the agent are stored in an experience replay buffer (state, action, reward-triplets). Random samples are taken from this buffer to train the ANN. This helps to reduce correlation between consecutive learning updates.
- **Target Network.** A separate target network with identical topology is used to generate Q-values for the training update. This makes the target more stable and therefore stabilizes the training process of the prediction network. Weights of the target network are updated with less frequency or gradually with a soft update strategy.
- **Fixed Replay buffer.** A certain percentage of the replay buffer may be kept fixed and is regularly shown to the ANN. This assures that the neural network does not forget learnings early in the training process.
- **Prioritized Experience Replay PER.** Experiences which incorporate a high TD error (difference between the predicted Q-value and the target Q-value) are shown to the ANN with a higher priority.
- **Boltzmann Exploration (Soft-max Exploration).** Boltzmann Exploration chooses the next action based on a soft-max distribution over the Q-values. The temperature of this distribution is decreased during the training process, enforcing a more and more deterministic policy. This mitigates the disadvantage of ε-greedy exploration which is, that it treats all non-greedy

actions equally when choosing randomly and thus disregarding better knowledge about the current value function. See Fig. 8.

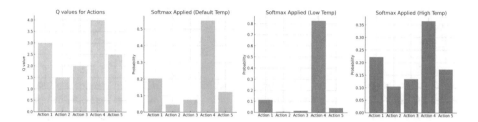

Fig. 8. Example of distributions for 5 actions for use in Boltzmann Exploration. The left diagram shows an example output of a Q-network which gives a value for every possible discrete action. The diagram with the green bars shows a soft-max function as applied to the Q-value from the first figure. This generates a probability distribution over the actions. The two diagrams on the right show this action probability distribution with a low and with a high temperature. A high temperature increases exploration, while a low temperature lets the agent behave more deterministically.

3.3   Policy Methods

Policy methods are a class of reinforcement learning algorithms that directly learn a policy, mapping states to actions. Two methods will be presented: Cross-Entropy Method and REINFORCE.

Cross-Entropy Method (CEM). The cross-entropy method (CEM) is a Monte Carlo, on-policy, model-free policy search method which iteratively refines the probability distribution over the space of possible policies. Policies are sampled and evaluated and the best ones are retained, improving the policy iteratively.

The algorithm is as follows:

1. **Initialize:** Randomly initialize the policy parameters.
2. **Generate Samples:** Collect trajectories using the current policy.
3. **Update Parameters:** Update the policy parameters using the elite trajectories (e.g. 30% top-performing samples).
4. **Repeat:** Iterate through steps 2 and 3 until convergence.

The ANN topology used for CEM is shown in Fig. 9 where also the training method is explained in more detail.

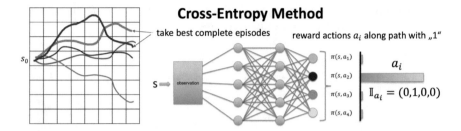

Fig. 9. This image depicts a policy network architecture utilized in the cross-entropy method CEM. The network takes an observation (state S) as input and outputs probabilities for each possible action. These probabilities $\pi(S, a_i)$ define the policy by indicating the likelihood of taking each action a_i given the current state. State-action pairs from top-performing trajectories are shown to the ANN. The respective actions a_i are then one-hot encoded and a multi-class cross-entropy loss function as in supervised learning is applied.

Monte Carlo Policy Gradient: REINFORCE (REward Increment = Non-negative Factor × Offset Reinforcement × Characteristic Eligibility) is a Monte Carlo, on-policy, model-free policy gradient method that directly maximizes the expected cumulative reward. Adding a baseline term helps reduce the variance of the gradient estimates.

Figure 10 illustrates the working of Monte Carlo policy gradient methods. Complete episodes are sampled, returns are received, and the according actions are rewarded with the respective rewards.

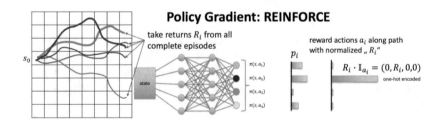

Fig. 10. REINFORCE: A policy gradient method which uses Monte Carlo methods.

Objective Function. The algorithm maximizes the expected cumulative reward by adjusting the parameters of the policy in the direction of greater reward. The *objective function* for REINFORCE, denoted $J(\theta)$, where θ represents the policy parameters (weights of the ANN), is typically the expected return:

$$J(\theta) = \mathbb{E}_{\tau \sim \pi_\theta} \left[\sum_{t=0}^{T} \gamma^t R(s_t, a_t) \right] \tag{25}$$

If the starting state s_0 is always the same, then $J(\theta) = V(s)$ (state-dependent objective), i.e. the objective function is the expected value in the starting state. If there is a probability distribution P_0 for the starting state, then $J(\theta) = \sum_s P_0(s) \cdot V(s)$ (state-agnostic objective).

REINFORCE uses gradient ascent on $J(\theta)$. The gradient of the objective function with respect to the policy parameters θ is estimated using complete (Monte Carlo) episodes generated by the policy π_θ (on-policy). The update rule for the policy parameters involves taking steps proportional to the product of the return and the gradient of the log-probability of the trajectory under the policy:

$$\theta \leftarrow \theta + \alpha \sum_{t=0}^{T} G_t \cdot \nabla_\theta \log \pi_\theta(a_t|s_t) \tag{26}$$

where G_t is the return from time step t and α is the learning rate. The Algorithm is as follows:

1. **Initialize:** Randomly initialize the policy parameters.
2. **Collect Trajectories:** Execute the policy to collect state-action-reward trajectories.
3. **Compute Returns:** Calculate the (accumulated discounted) returns G_t for each time step.
4. **Update Policy:** Update the policy parameters according to (26).
5. **Repeat:** Iterate through steps 2–4 until convergence.

The policy gradient theorem shows how the update formula (26) is derived.

Policy Gradient Theorem: Given a policy π_θ parameterized by θ, the gradient of the expected return $J(\theta)$ with respect to the policy parameters is:

$$\nabla_\theta J(\theta) = \mathbb{E}_{\tau \sim \pi_\theta} \left[\sum_{t=0}^{T} \nabla_\theta \log \pi_\theta(a_t|s_t) G_t \right]$$

where G_t is the return following time t, $\pi_\theta(a_t|s_t)$ is the probability of taking action a_t in state s_t under policy π, and τ represents a trajectory of states and actions.

Proof: It has to be shown that

$$\nabla_\theta J(\theta) = \nabla_\theta \mathbb{E}_{\tau \sim \pi_\theta} \left[\sum_{t=0}^{T} \gamma^t R(s_t, a_t) \right] \overset{!}{=} \mathbb{E}_{\tau \sim \pi_\theta} \left[\sum_{t=0}^{T} \nabla_\theta \log \pi_\theta(a_t|s_t) \cdot G_t \right] \tag{27}$$

This means that the gradient of the expected return (with respect to the weights θ of the policy) is the expected value of the log-probabilities of the taken actions and the cumulative reward. In other words: The gradient of the objective function can be expressed as the gradient of local values of the policy $\nabla_\theta \log \pi_\theta(a_t|s_t)$, which in practice can be calculated with ML-optimizers like tensorflow/Keras, and the total return G_t which is determined empirically. The proof is composed of two steps.

Step 1: According to (12) the objective function $J(\theta)$ can be written as the expected total return:

$$J(\theta) = \mathbb{E}_{\tau \sim \pi}[G(\tau)] = \sum_{\tau} P(\tau|\theta) \cdot G(\tau)$$

where $P(\tau|\theta)$ is the probability of trajectory τ under policy π according to the *policy induced trajectory distribution* (11), and $G(\tau)$ is the total return for trajectory τ. To find the gradient of $J(\theta)$, it has to be differentiated with respect to θ:

$$\nabla_\theta J(\theta) = \nabla_\theta \sum_{\tau} P(\tau|\theta) \cdot G(\tau) = \sum_{\tau} \nabla_\theta P(\tau|\theta) \cdot G(\tau)$$

The chain rule for derivatives implies $\ln(f(x))' = f'(x) \cdot \frac{1}{f(x)}$. Applying this to $\nabla_\theta P$ gives $\nabla_\theta P = P \cdot \nabla_\theta \log P$, and thus:

$$\nabla_\theta J(\theta) = \sum_{\tau} P(\tau|\theta) \cdot \nabla_\theta \log P(\tau|\theta) \cdot G(\tau)$$

Using the definition of the expected value this gives

$$\nabla_\theta J(\theta) = \mathbb{E}_{\tau \sim \pi_\theta} [G(\tau) \cdot \nabla_\theta \log P(\tau|\theta)] \tag{28}$$

Step 2: Now the policy induced trajectory distribution (11) is substituted for $P(\tau|\theta)$

$$\nabla_\theta J(\theta) = \mathbb{E}_{\tau \sim \pi_\theta} \left[G(\tau) \cdot \nabla_\theta \log \prod_{t=0}^{T-1} \pi(a_t|s_t) \cdot P(s_{t+1}|s_t, a_t) \right]$$

Note that $G(\tau)$ is independent of θ and can therefore be written in front of the gradient. Now the product rule for logarithm $\ln(a \cdot b) = \ln a + \ln b$ is applied to the product to receive

$$\nabla_\theta J(\theta) = \mathbb{E}_{\tau \sim \pi_\theta} \left[G(\tau) \cdot \sum_{t=0}^{T-1} (\nabla_\theta \log \pi(a_t|s_t) + \nabla_\theta \log P(s_{t+1}|s_t, a_t)) \right]$$

Since the transition probabilities $P(s_{t+1}|s_t, a_t)$ are generated by the system and are not dependent on θ, the gradient is zero. Therefore

$$\nabla_\theta J(\theta) = \mathbb{E}_{\tau \sim \pi_\theta} \left[\sum_{t=0}^{T-1} G_t \cdot \nabla_\theta \log \pi_\theta(a_t|s_t) \right] \tag{29}$$

∎

The policy gradient theorem shows, that the gradient of the policy depends on the log-gradient of the local policy values.

Implementing REINFORCE with Cross-Entropy Loss. The multi-class cross-entropy loss in a typical classification problem for classes c is given by:

$$L(\theta) = - \sum_c \mathbb{I}_{\bar{c}}(c) \cdot \log(p_c) = - \log(p_{\bar{c}}) \tag{30}$$

where $\mathbb{I}_{\bar{c}}(.)$ is the one-hot encoded vector for the true class $\bar{c}$, p_c is the predicted probability for class c. In order to be able to use ML-optimizers for REINFORCE, the *label* distribution $\mathbb{Y}$ has to be a probability distribution. This is achieved by creating a label probability distribution $\mathbb{Y}(.)$ for the respective state s_t by *pulling* the predicted probability distribution $\pi_\theta(.|s_t)$ with value G_t towards the action $\bar{a}$ which was taken in the episode

$$\mathbb{Y}_{s_t}(.) = (1 - G_t) \cdot \pi_\theta(.|s_t) + G_t \cdot \mathbb{I}_{\bar{a}}(.) \tag{31}$$

This requires that the returns G_t for all states from the sampled episodes have been normalized to $G \leftarrow \frac{G - \mu}{\sigma}$. If the normalized $G_t = 0$, then $\mathbb{Y}_{s_t}(.) = \pi_\theta(.|s_t)$ and no training occurs. If the normalized $G_t > 0$, then the respective action is rewarded, if $G_t < 0$ then the respective action is penalized.

It can be shown that applying multi-class cross-entropy loss from (30) when using the label probability distribution $\mathbb{Y}$ from (31) leads to the gradient as calculated in Eq. (29), when calculated for the output neurons of the policy network before applying the soft-max activation function.

3.4   Actor-Critic Methods

Due to high variance of Monte Carlo methods, REINFORCE generally doesn't have very good convergence properties. Figure 11 shows typical results from training sessions with identical hyperparameters for the cart pole problem in OpenAI gym.

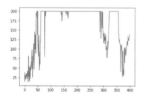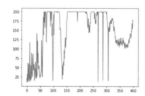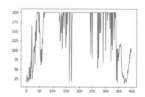

Fig. 11. Learning curves from training episodes using identical hyperparameters. The differences are due to stochasticity in the system and the policy. It can be seen, that the agent keeps forgetting its progress.

Actor-Critic methods combine policy-based (Actor) and value-based (Critic) reinforcement learning approaches to optimize a policy. This introduces the concept of learning in temporal differences (bootstrapping) for policy gradient methods. As a consequence, variance will be reduced. This section will cover

Advantage Critic, and Proximal Policy Optimization (PPO) as examples for Actor-Critic algorithms.

The following two equations illustrate the differences between an Actor-Critic (AC) and a Monte Carlo Policy Gradient (MC-PG) method. See the policy gradient theorem and Eq. (28):

$$\text{Actor-Critic: } \nabla_\theta J(\theta) = \mathbb{E}_{\tau \sim \pi_\theta} [A^\pi \cdot \nabla_\theta \log P(\tau|\theta)]$$
$$\text{MC-PG: } \nabla_\theta J(\theta) = \mathbb{E}_{\tau \sim \pi_\theta} [G(\tau) \cdot \nabla_\theta \log P(\tau|\theta)]$$

It can be seen that the difference lies in the factor in front of the gradient-logarithm. While MC-PG methods use the total return $G(\tau)$ which has a high variance, AC methods use a local *advantage* A^π as explained in the following section:

Advantage Actor-Critic A2C. In the Advantage Actor-Critic (A2C) algorithm, the Critic estimates the advantage function $(A(s,a))$, representing the advantage of taking action a in state s over the baseline value. The definition of the advantage is

$$A^\pi(s_t, a_t) := Q^\pi(s_t, a_t) - V^\pi(s_t) \tag{32}$$

The advantage A^π gives the relative quality of each action.

Equation (33) shows the gradient of the actor-critic method, based on the local action probabilities provided by the policy π_θ as derived in Step 2 of the proof of the policy gradient theorem.

$$\nabla_\theta J(\theta) = \mathbb{E}_{\pi_\theta} \left[\sum_{t=0}^{T-1} \nabla_\theta \log \pi_\theta(a_t|s_t) \cdot A^{\pi_\theta}(s_t, a_t) \right] \tag{33}$$

The algorithm has the same basic form as the Monte Carlo Policy Gradient method. However, step 3 is different in that the advantage is calculated.

Algorithm:

1. **Initialize:** Randomly initialize the policy (Actor) and the advantage function (Critic) parameters.
2. **Collect Data:** Interact with the environment to collect state-action-reward tuples.
3. **Update Advantage Function:** Minimize the advantage function's error to update the Critic.
4. **Update Policy:** Maximize the advantage values to update the policy.
5. **Repeat:** Iterate through steps 2–4 until convergence.

As can be seen in Fig. 12a and 12b, the state value function V is approximated by the neural network. In order to obtain A^π, the Q-value needs to be calculated. This is done using the Bellman Principle (18) such that

$$Q(s, a) \leftarrow r + \gamma \cdot V(s')$$

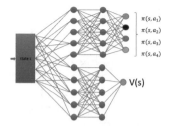

(a) Actor-critic with shared network.   (b) Actor-critic with separate networks.

Fig. 12. Different architectures for Actor-Critic Networks. a: The shared network architecture has less parameters and may therefore learn faster. However, learning may be unstable, because the policy and the value function have different gradients. b: The separate network architecture learns slower because V depends on the policy π which therefore needs to be learned first. In both cases, the loss function of one part of the ANN (e.g. the part for the value approximation) can be scaled with a parameter to balance the two networks.

where r is the immediate reward. s' is obtained from the policy network when applying action a: $s' \sim \pi(a|s)$. Substituting into the definition of the advantage gives

$$A(s, a) = r + \gamma \cdot V(s') - V(s).$$

Generalized Advantage Estimation (GAE). Generalized Advantage Estimation (GAE) as introduced in [6] provides a way to estimate the advantage function with a balance between bias and variance. The formulation is as follows:

TD Residual: The Temporal Difference (TD) residual at each time step is defined as:

$$\delta_t = r_t + \gamma V(s_{t+1}) - V(s_t)$$

Here, r_t is the reward at time t, $V(s)$ is the value function at state s, and γ is the discount factor.

GAE Calculation: GAE is computed using these TD residuals, incorporating a decay parameter λ to adjust the bias-variance trade-off:

$$\hat{A}_t^{GAE} = \sum_{k=0}^{T-t-1} (\gamma\lambda)^k \delta_{t+k}$$

This equation represents a weighted sum of k-step TD residuals. Figure 13c illustrates different weightings for different path lengths.

The parameters γ and λ in GAE control the trade-off between bias and variance. A lower λ value results in estimates with less variance but more bias, while a higher λ achieves the opposite. Adjusting λ allows GAE to provide a flexible approach to estimate the advantage function, facilitating more stable and efficient updates in reinforcement learning algorithms.

Proximal Policy Optimization (PPO). The problem with A2C (Advantage Actor-Critic) and similar on-policy algorithms is that they can be sample inefficient and sensitive to the choice of hyperparameters. They often require a large number of samples to learn effectively because they can only use current policy samples to improve the policy and cannot reuse data from previous policies (due to being on-policy). Moreover, their performance can drastically change with different step sizes for updating the policy. Figure 13a illustrates the problem of finding an optimal policy.

The trick in PPO (Proximal Policy Optimization) (see [7]) is to address these issues by using a clipped objective function as illustrated in Fig. 13b, which prevents the policy from changing too much at each update. It allows for using a larger step size while still maintaining stable and robust learning. PPO maintains the benefits of on-policy learning but with a more stable and reliable learning process.

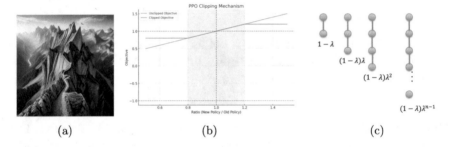

(a)   (b)   (c)

Fig. 13. Finding a maximum in a policy requires to follow a narrow path (a) in policy space. If an update step is taken which is too large, the agent *forgets* or drops off its path. This results in a learning progress as shown in Fig. 11. b: Illustration of the PPO Clipping Mechanism: Balancing Exploration and Exploitation in Reinforcement Learning. c: In Generalized Advantage Estimation (GAE) there is a balance between bias and variance.

In PPO, a policy improvement ratio is introduced, denoted by

$$r_t(\theta) = \frac{\pi_\theta(a_t|s_t)}{\pi_{\theta_{\text{old}}}(a_t|s_t)}$$

r_t indicates how much the new policy differs from the old policy.

The **Objective Function** makes use of a clip function as shown in 13b to decide weather to update the policy

$$L(\theta) = \mathbb{E}\left[\min\left(r_t(\theta)A_t, \text{clip}(r_t(\theta), 1 - \epsilon, 1 + \epsilon)A_t\right)\right]$$

where A_t is the advantage function.

The **Algorithm** is similar to A2C, with the exception that the mentioned objective function is used. Exploration is usually done by choosing an action randomly given the current probability distribution of the prediction in the respective state.

The objective function of PPO prevents the policy from changing too much in a single update. The clipping mechanism limits the size of the policy update. This allows PPO to reuse data from multiple epochs which makes PPO more sample efficient, more stable and less sensitive to hyperparameters than A2C.

3.5   Off-Policy Deep RL and Continuous Control

In this chapter, off-policy reinforcement learning methods that are effective for continuous control tasks are discussed. There will be a particular focus on SAC (Soft Actor-Critic) and DDPG (Deep Deterministic Policy Gradient). These methods are distinguished by their ability to learn from experiences generated by a policy different from the one being improved (off-policy). They utilize a replay buffer, which stores past interactions with the environment and allows the algorithms to learn from this stored data multiple times.

For discrete action spaces, a neural network might use separate output neurons for each possible action, a technique commonly seen in methods like DQN. However, when dealing with continuous action spaces, SAC and DDPG use a different approach. They produce real-valued outputs that correspond to the specifics of the action space. An example for a continuous control is the steering angle of an autonomous vehicle.

DDPG is known for creating a deterministic policy, directly linking states to specific actions. This is in contrast to SAC, which creates a stochastic policy by outputting parameters that define a probability distribution, such as a Gaussian. This stochastic approach helps maintain a balance between trying out new actions (exploration) and using actions known to yield good results (exploitation).

It should be noted that PPO, as outlined in Sect. 3.4, can be adapted for continuous stochastic control tasks. This adaptation involves implementing a policy network that outputs parameters-specifically, a mean μ and standard deviation σ-for the action distribution.

Both SAC and DDPG are made to work with continuous action spaces and make good use of the off-policy learning approach.

Soft Actor-Critic SAC. SAC incorporates elements from DQN, like the use of two separate networks for Q-value estimation, enhancing stability in value function approximation. This dual-network approach, combined with direct policy learning, positions SAC as an off-policy method, enabling learning from experiences generated by various policies. Unlike DQN, SAC's architecture is tailored for continuous action spaces.

Soft Actor-Critic, as introduced in [8], integrates entropy regularization into its objective function. This encourages the policy to maintain a level of exploration by preferring higher entropy policies, preventing premature convergence to overly deterministic behavior. The difference to the method described in Fig. 8

is that Boltzmann exploration directly uses the value estimates to form a probability distribution for action selection while entropy regularization integrates exploration into the learning objective itself.

While the standard RL methods maximize the expected sum of rewards $\sum \mathbb{E}_\pi[r(s_t, a_t)]$ (dropping the discount factor γ for clarity), in Soft Actor-Critic the expected entropy of the policy is included in the objective function, given by:

$$J(\theta) = \mathbb{E}_\pi \left[\sum_{t=0}^{T} (r(s_t, a_t) + \alpha \cdot \mathcal{H}(\pi(\cdot|s_t))) \right] \tag{34}$$

Where $\mathcal{H}(\pi(\cdot|s_t))$ is the entropy of the policy and α is a temperature parameter that balances the importance of the reward and entropy terms.

The entropy term is defined as

$$\mathcal{H}(\pi(\cdot|s_t)) := -\sum_{a \in A} \pi(a|s_t) \log(\pi(a|s_t)). \tag{35}$$

The policy network (actor) of SAC outputs a probability distribution for actions, typically a Gaussian distribution with parameters like mean and standard deviation. Actions are sampled from this distribution.

$$a \sim \pi(.|s) = \mathcal{N}(\mu(s), \sigma(s)) \tag{36}$$

where $\mathcal{N}$ denotes the Gaussian distribution with mean μ and variance σ.

The particular design of the policy network enables SAC to be used for continuous action spaces.

Soft Actor-Critic uses several neural networks to approximate

- the soft state value function $V_\psi(s)$
- the soft Q-function $Q_\theta(s, a)$
- the policy $\pi_\phi(a|s)$

where ψ, θ, ϕ denote the network parameters.

The soft state value function is defined as

$$V(s) := \mathbb{E}_{a \sim \pi}[Q(s, a) - \alpha \cdot \log \pi(a|s)] \text{ definition of soft } V$$
$$= \sum \pi(a|s) \cdot [Q(s, a) - \alpha \cdot \log \pi(a|s)] \text{ definition expected value}$$
$$= \sum \pi(a|s) \cdot Q(s, a) + \alpha \cdot \mathcal{H}(\pi(\cdot|s_t)) \text{ definition of entropy } \mathcal{H}$$

This term encourages to select high-reward actions (as indicated by the Q-value) with the tendency to maintain diversity in action selection (as encouraged by the entropy in $-\mathbb{E} \log \pi$). The inclusion of the negative log probability of the chosen action encourages the policy to favor actions that are not only high in expected return but also high in entropy, promoting exploration.

The soft Q-function

$$Q_\theta(s_t, a_t) = r(s_t, a_t) + \gamma \mathbb{E}_\pi[V_\psi(s_{t+1})] \tag{37}$$

is updated using the soft state value function V_ψ. Then the policy is updated using the soft Q-function. The objective function for the policy is given by

$$J_\pi(\phi) = \mathbb{E}\left[\alpha \cdot \log(\pi_\theta(a_t|s_t)) - Q_\phi(s_t, a_t)\right] \tag{38}$$

also favouring a probability distribution with high entropy.

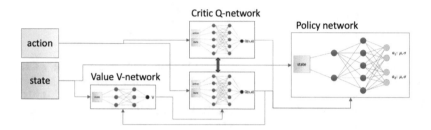

Fig. 14. This illustration shows the neural network architecture of the Soft Actor-Critic (SAC) algorithm. The architecture comprises three networks: the network for the state value function; the Q-network (Critic), consisting of two separate networks that estimate the Q-value function Q(s,a), and the Policy network, which outputs a probability distribution over actions, parameterized by mean μ and standard deviation σ, given the state. This structure enables SAC to balance exploration and exploitation in continuous action spaces.

Figure 14 illustrates the neural network structure of Soft Actor-Critic.

Deep Deterministic Policy Gradients (DDPG). Deep Deterministic Policy Gradients (DDPG) as described in [9] is an actor-critic algorithm designed for continuous action spaces. It combines ideas from both value-based methods and policy-based methods, leveraging a deterministic policy and a Q-function. It is particularly useful in robotic applications.

Figure 15 illustrates the fundamental architecture of DDPG, an actor-critic method. The policy network (actor) deterministically outputs continuous action values for each state. The policy network is trained using a policy gradient approach, where the training involves a Q-value provided by a separate network, known as the critic. This Q-network concept shares similarities with DQN, as detailed in Sect. 3.2. In DDPG, there are two sets of networks: the actor-critic networks and their corresponding target networks, which are softly updated to enhance training stability. The critic network takes the current state and action as inputs, along with data sampled from the replay buffer, and outputs a Q-value through a single linear neuron. This Q-value represents the estimated value of taking the given action in the current state.

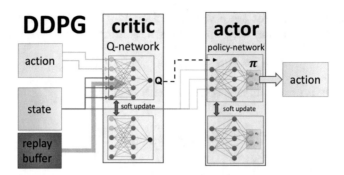

Fig. 15. Neural Network architecture involved in Deep Deterministic Policy Gradient. The state, the previously taken action, and data from a replay buffer are taken as input to a Q-network. The Q-value is taken to update the policy network. Both neural networks are copied and use soft target updates for stability. The output of the policy network is a continuous deterministic action.

3.6   Conclusion and Summary

The article provided an introduction to reinforcement learning, describing some of the most recent model-free methods like PPO, SAC, and DDPG. These algorithms have spawned various versions, enhancements, and practical implementations, such as the ML-agents framework in the Unity game engine.

Current research in reinforcement learning is expanding towards broader applications, though real-world problems pose significant challenges. One major difficulty is the slow acquisition of real-world data and the constraints on exploration in real-world settings. Off-line reinforcement learning addresses learning effective policies from fixed datasets without additional online interaction with the environment. This approach is particularly relevant for scenarios where real-time exploration is either impractical or risky.

Another area of interest are large and complex problems characterized by sparse rewards, where feedback is minimal and infrequent. Potential solutions may involve integrating generative AI, Large Language Models (LLMs), or transformers to better handle these challenges. Additionally, there is ongoing work on developing foundational models for robotics, aimed at equipping algorithms with the necessary tools to address complex, multifaceted problems.

As reinforcement learning continues to evolve, its application to real-world scenarios becomes increasingly feasible, opening up new avenues for innovation and problem-solving.

References

1. Sutton, R., Barto, A.: Reinforcement Learning: An Introduction. MIT Press, Cambridge (2018)
2. Lapan, M.: Deep Reinforcement Learning Hands-On: Apply Modern RL Methods, with Deep Q-Networks, Value Iteration, Policy Gradients, TRPO, AlphaGo Zero and More. Packt Publishing, Birmingham (2018)

3. Palmas, A., et al.: The Reinforcement Learning Workshop: Learn How to Apply Cutting-Edge Reinforcement Learning Algorithms to a Wide Range of Control Problems. Packt Publishing Ltd., Birmingham (2020)
4. Graesser, L., Keng, W.: Foundations of Deep Reinforcement Learning. Addison-Wesley Professional, Boston (2019)
5. Mnih, V., et al.: Human-level control through deep reinforcement learning. Nature **518**, 529–533 (2015). http://dx.doi.org/10.1038/nature14236
6. Schulman, J., Moritz, P., Levine, S., Jordan, M., Abbeel, P.: High-dimensional continuous control using generalized advantage estimation. arXiv Preprint arXiv:1506.02438 (2015)
7. Schulman, J., Wolski, F., Dhariwal, P., Radford, A., Klimov, O.: Proximal policy optimization algorithms. arXiv Preprint arXiv:1707.06347 (2017)
8. Haarnoja, T., Zhou, A., Abbeel, P., Levine, S.: Soft actor-critic: off-policy maximum entropy deep reinforcement learning with a stochastic actor. In: International Conference on Machine Learning, pp. 1861–1870 (2018)
9. Lillicrap, T., et al.: Continuous control with deep reinforcement learning. In: International Conference on Learning Representations, Puerto Rico (2016)

Autonomous Emergency Landing of an Aircraft in Case of Total Engine-Out

Steffen Flämig$^{(\boxtimes)}$ and Wolfram Schiffmann

FernUniversität in Hagen, Unistr. 1, 58084 Hagen, Germany
steffen@flaemig.de, wolfram.schiffmann@fernuni-hagen.de

Abstract. An engine failure is the nightmare of every pilot. In order not to give away the best options for an emergency landing, a decision on the further flight route in gliding must be made as soon as possible. It is important to divide the remaining height available when deciding on an emergency landing in such a way that the runway threshold is still reached at a suitable flaring height despite the influence of wind. This route planning can be carried out with the help of smart emergency landing assistance systems in a very short time and with high accuracy. The present article describes the solutions developed so far by the SmartFly-Solutions team.

1   Introduction

What happens when the engine (or engines) of an airplane fail? Fortunately, the airplane doesn't immediately fall down from the sky but it becomes a sailplane. Especially fixed wing aircrafts can efficiently transform the potential energy from the altitude above the runway into kinetic energy for moving. E.g. a typical airliner can glide nearly 200 km if the engine failure happens 10.000 m above ground level. However, height is constantly being consumed. Thus, one must choose a smart route that, with this residual height, brings the plane exactly to the beginning of the runway. Moreover, the aircraft must arrive at the right altitude and heading for the landing which starts over the so-called threshold. The chosen flight route must also compensate for the influence of wind. While this is easy with a working engine, it is a big challenge when the engine fails. If the route doesn't fit or the pilot doesn't follow it exactly, you either end up too short or too long regarding the available runway. This means that you touch down before the runway or that you overshoot (see Fig. 1).

In the Fig. 1 you see three engine-out approaches that start at the same position. In all cases the routes are of the same length which is equal to the excess altitude over the runway times the glide ratio of the aircraft. If the pilot immediately turns towards the runway the aircraft will probably overshot the runway, e.g. it will come too long. If on the other hand, the pilot flies too long into the downwind direction before he turns in, the aircraft will not reach the threshold of the runway, it will come too short. In contrast, a successful landing route

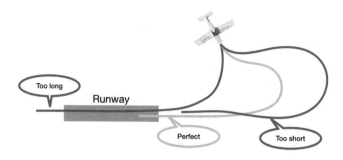

Fig. 1. Three possible emergency approaches

lies somewhere in between those extremes. The pilot or better an appropriate assistance software divides the available glide length in such a way that it ends close to the runway's threshold.

The most prominent example for total engine-off emergency happened in January 2009 shortly after the takeoff of an Airbus 320 in La Guardia. Captains Sullenberger's plane collided at just 2700 ft with a swarm of geese and both engines were knocked out. From this height Sullenberger had no more than approx. 35 s to make a decision to return to LaGuardia which would have been possible. Due to the surprise effect and the pilots' uncertainty as to whether a return would be possible, this period passed unused. Eventually, as we today know, Sullenberger decided to make an emergency landing in the Hudson River, which he completed with flying colors. But, it could have ended badly, because water landings are very problematic, as the airplane can overturn and break. This can be triggered by even the slightest movement of water or wind when ditching.

The rest of the paper is organized as follows: In the next section we describe the popular high-key/low-key heuristic to support pilots in case of an engine failure. Than, we describe the so called glide range ring as basic approach of commercial computer-based avionics systems. Unfortunately, those applications don't provide a trajectory from the emergency position to the landing area of a runway. In the main part of our contribution we describe our application Safe2Land to that solves this problem by using a fine grained model of so-called kinematoide chains. Safe2Land compensated the influence of the wind and can even command an autopilot to fly the aircraft autonomously to the runway.

2   High-Key/Low-Key Heuristics

The most frequently described method is the "high key - low key" technique (see Fig. 2). It essentially begins at the high-key, approximately 2,500 ft above ground level (AGL) abeam to the selected touchdown point. The pilot should then fly with a 360° descent turn towards this touchdown point. It first flies around the it - in a left-hand semicircle - and arrives at the end of a "downwind approach" at approximately 1,500 ft AGL that is called low-key. From there she flies towards the runway threshold in a more or less curved 180° left turn. The curvature depends on the wind and the remaining AGL. Even if the procedure outlined provides a certain orientation, it of course assumes that the above-mentioned AGL height abeam to the threshold can actually be reached and hit reasonably accurately. The term abeam is also not precisely defined numerically. Moreover, altitudes and configurations at the key points are determined by the aerodynamic characteristics of the aircrafts and can change in different publications [1]. However, to hit the high-key a similar landing planning is required as for the actual emergency landing. The term transverse is also not precisely defined numerically. However, pressing the high button requires similar landing planning to the actual emergency landing. Actually, with the high-key/low-key method you simply postpone the problem to an earlier point in time in order to then have a little more leeway for corrections.

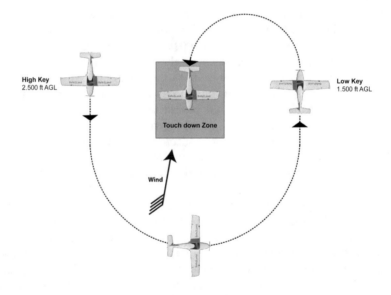

Fig. 2. Principle of the High-key/Low-key procedure

3   Computer-Based Avionics Systems

Besides the above described heuristic there are also computer-aided methods to solve the above emergency landing problem? To assist the pilot in determining whether an airport can be reached, most of today's navigation applications offer a glide range ring. Examples of this can be found at ForeFlight, SkyDemon,

Garmin Pilot etc. Even if it is the most advanced commercial solution, Garmin's SmartGlide remains still a pretty simple system [2]. In the event of an engine failure, it checks all publicly registered airfields within the gliding range and selects the most suitable airfield for the aircraft. It then brings the aircraft via autopilot (in gliding flight) to a sufficient altitude above this airfield, from where the pilot has to carry out the actual landing procedure to the threshold. Similar to the other mentioned avionic applications the pilot gets no concrete route from the current position to the threshold of the selected airport's runway.

To compute the glide range ring one considers the aircraft position as the center and computes the loss of altitude along a distance d over a number of equidistant radials from that center. Starting from the aircrafts heading an increment of $360/n$ is added $1..(n-1)$ times in order to cover the whole glide range ring. The altitude loss is calculated by the distance d from the center divided by the wind-corrected glide ratio g of the aircraft type to

$$\Delta h = d/E \tag{1}$$

The wind-corrected glide ratio d depends on whether a head or tail wind is blowing along the concerned radial. If there is a head wind component w_h, the best glide speed of the aircraft is denoted by v_{bg} and the corresponding sink rate is denoted by v_s the wind correct glide ratio will become

$$E = \frac{v_{bg} - w_h}{v_s} \tag{2}$$

Supposed one only knows v_{bg} and the windless glide ratio is E_0 one can substitute v_s by $\frac{v_{bg}}{E_0}$ and one gets

$$E = E_0 \cdot \left(1 - \frac{w_h}{v_{bg}}\right) \tag{3}$$

Similarly, in case of tail wind w_t one gets

$$E = E_0 \cdot \left(1 + \frac{w_t}{v_{bg}}\right) \tag{4}$$

Supposed the aircraft altitude at the center is denoted by h one calculates the remaining altitude as $h - \Delta h$ as a function of d using the wind-corrected glide ratio E. Eventually, for d_R the altitude h will hit the surface of the surrounding terrain and thus we get the length of the corresponding radial. If we do this computations for all the radials we can draw the glide range ring.

Although in this way both terrain profile and wind are taken into account, it is incorrectly assumed that changes in aircraft direction (necessary to approach the runway) don't increase the required glide path length. In the Fig. 3) a simplified glide range ring is shown together with several runways. Neither wind or nor terrain distortion is considered for simplicity. Thus, it looks like a perfect circle. As you can see the available glide length is just sufficient to make the threshold of the three runways (black rectangles) that are located mainly in the direction of the aircraft's heading. The glide footprint considers the reduction of range due to directional changes. Although the Figure makes it seem that way, the other five runways (filled with grey) could not reached for landing the aircraft.

To wrap up, some runways shown within the glide range are not landable for fixed-wing aircraft. If the emergency landing targets are opposite or perpendicular to the current flight direction, turns must be flown, which reduces the original gliding range. Thus, it may happen that an emergency landing site selected by the pilot cannot be reached because the glide area ring shrinks due to changes in direction. The loss of altitude due to directional changes within the gliding area is therefore not taken into account at all, i.e. the actual footprint of the area around the aircraft that can be reached during gliding actually looks completely different than commercial avionics systems suggest (see [3]).

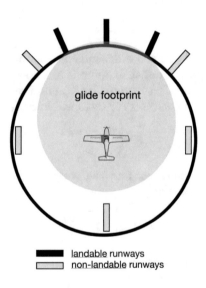

Fig. 3. Landable vs. non-landable runways displayed with a glide range ring (no wind or terrain) and the real footprint due to directional movements

Another Garmin avionics product called "Autoland" is frequently mentioned and claimed to already implement the same emergency functionality as our Safe2Land. However, this system only helps in case of pilot incapacitation. Garmin's Autoland still requires the engine to be working and enough fuel on board to make a certified RNAV approach to the next registered airport with an certified RNAV approach. In contrast, Safe2Land can handle pilot and engine failures simultaneously and doesn't rely on the RNAV approach restriction.

4 Optimized Emergency Approach Routes that Provide Altitude Division

In the following we first analyse the general problem, show how the wind influence is compensated, sketch our Emergency Landing Field Identification method (ELFI) and introduce the first version of our Safe2Land application. Thereafter, we introduce the current and more elaborated version of it.

4.1 General Problem Analysis

At first, we look for methods to calculate flyable route to the threshold. Several opportunities for route calculation will be described and illustrated below. We consider a fixed-wing aircraft and assume that the destination for the emergency landing is already known. For the sake of simplicity, we will start with the windless case and first show how to reach the runway from an arbitrary emergency position within reach. This position includes the geographical location consisting of longitude, latitude, altitude (above sea level) and the heading of the aircraft at this point.

We also assume a sufficient altitude to reach the threshold of the runway and consider the necessary directional changes. While flying with the best glide speed, a pilot can change the heading by turning left or right and in this way fly from any emergency position towards the center of the emergency landing field. However, for landing she must bring the aircraft to an appropriate position of the final approach. From there she must be able gliding to somewhere between the threshold and the center of the runway. Thus, generally the route must contain at least two circular segments that are tangentially connected by at least one section of straight flight. If we glide along the route shown in Fig. 4, the excess altitude above ground is converted into distance (more precisely, potential energy into kinetic energy). This means that the altitude is reduced by the glide and, ideally, the aircraft arrives at an appropriate height above the threshold for hovering until its speed is reduced sufficiently. Then the airplane stops flying (stalls) and settles down on the runway.

To model the descent, we need the glide ratio and the speed of the best glide, which the pilot must of course adhere to as precisely as possible. Since the route also contains turns, the best glide speeds and glide numbers for the attitudes flown must also be known in order to correctly model the flight behavior. Based on the exact GPS position and these parameters, the approach route can be determined using suitable algorithms so that the pilot arrives perfectly for hovering when he reaches the threshold.

In practice, one has to consider four different flight patterns for each runway direction (corresponding to the combinations of the two possible turn directions left-left, left-right, right-left, right-right), see also Fig. 4. More details about those three dimensional Dubins paths are given in [4] and [5].

{LSL, RSR, LSR, RSL}

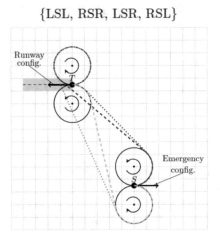

Fig. 4. Modified Dubins path from emergency position to the runway's threshold

4.2   Compensation of Wind Influence

To simplify things, we assumed in the last section that there was no wind. If, as usual for humans, one assumes an earth-fixed reference system, the orbit calculation becomes quite complicated due to the additional influence of wind. If you look at an aircraft flying in a circle from the ground, you will only see a circle when there is no wind. A non-vanishing wind deforms the circle into a so-called trochoid. Figure 5 illustrates how the calculation of these trochoids complicates the calculation of optimized approaches. This also increases the route calculation time significantly. However, this problem can be solved by moving the reference point into the cockpit and also virtually moving the target runway into the wind. If you also know how long the plane will sink you can easily calculate the displacement vector. Fortunately, the sink rate is known via the velocity polar and therefore the time of glide can be determined. If the aircraft now flies to this virtual runway as in a calm situation, the wind will blow it exactly to the real runway threshold. With this trick (also see [4]), which was developed by our team, you can apply any approach procedure without any changes in windy situations. Our method, called moving target principle, can even manage difficult wind profiles with varying wind vectors in different altitude layers (Fig. 6).

4.3   Database of Emergency Landing Fields

If the flight altitude above ground is too low in the emergency position, it may be that no registered airport is within reach for an emergency landing. Without an emergency landing assistance system, the pilot then has to look for a suitable emergency landing field in the area. Unfortunately, the field of vision from the cockpit is very limited because the panel obscures the downward view and the pilot can actually only see sideways to the left or right in an angle range between

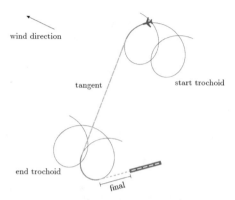

Fig. 5. Three dimensional LSL-Dubins path from emergency position to an extended final to consume the total available glide length

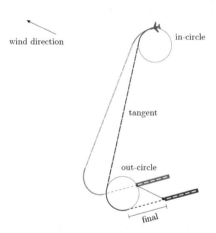

Fig. 6. Illustration of the moving target principle

30–90°. This means that only about a third of the area around the aircraft can be observed. In addition, the length and quality of possible emergency landing areas can hardly be estimated systematically due to looking diagonally downwards and the high time pressure.

For these reasons, our team developed new methods for the automated identification of emergency landing fields using freely available geodata. A database with emergency landing fields for small aircraft could be generated for various federal states (such as North Rhine-Westphalia and Saxony). An example from North Rhine-Westphalia is shown in Fig. 7. More than 100,000 emergency landing fields with minimum lengths of 450 m (green) and 300 m (red) could be identified in the whole county. You can imagine how important such a database becomes in regions with a small number of emergency landing options (e.g. in the Sauerland).

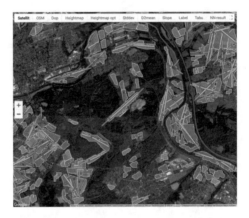

Fig. 7. Identified emergency landing fields around the airport Arnsberg (EDLA)

4.4   Safe2Land App (Version 1)

The trajectory calculations described above (based on Dubins paths) and the database of emergency landing fields have been integrated into an Android app Safe2Land. This app was initially successfully tested in the flight simulator and later on in numerous real test flights with a wide variety of small aircraft. Tests have even been carried out in which Safe2Land's flight guidance was carried out directly by an autopilot. The real tests were conducted with a DA42 (research aircraft from the Technical University of Munich) both at the DLR in Oberpfaffenhofen and at the airfield in Siegerland (EDGS).

Our research and real world tests on the Safe2Land emergency landing assistance system have shown that the app provides pilots with the best possible support in emergencies and that it even reduces heart rate and stress levels during real emergency landings. We are currently in the process of carrying out more detailed test scenarios in the flight simulator with psychologists and aviation physicians while comparing our procedures with other computerized solutions.

In addition to Safe2Land's Dubins paths, we have now developed an even more powerful method called "Smart Engine-off Emergency Advisor (SEA)", which is combined with so-called kinematoid chains and which models the atmospheric conditions at different altitudes more precisely. Safe2Land Version 2 is based on SEA that generates a list of reference points that can be calculated very efficiently. A kinematoide chain of up to 1000 segments is then constructed along these reference points. The fundamentals of Safe2Land Version 2 will be presented in the next part of this paper.

4.5   Safe2Land App (Version 2)

The Limits of Previous Approaches. The methods described so far for calculating gliding trajectories are based on some simplifications. These lead to errors, especially in complex conditions or long trajectories. One can get these errors under control by constantly recalculating the trajectory during the flight. This reduces the potential for errors because the part of the entire flight route that is still in front of the aircraft at any given time becomes shorter and shorter. However, this requires that sufficient reserves are always available for correction. These must have been planned accordingly in advance. It would be better if the trajectory is modeled from the beginning with less simplifying assumptions.

When gliding, a powerless aircraft loses altitude almost permanently. The density of the air surrounding the aircraft also increases continuously. The gliding characteristics of the aircraft, essentially determined by lift (5) and drag (6), in turn depend on the density of the surrounding air, the relative speed to it as well as the aerodynamic parameters of the aircraft. The latter can essentially be described by the glide ratio E (1).

$$Lift = A \times c_a \times \frac{\rho}{2} \times v^2 \tag{5}$$

$$Drag = A \times c_w \times \frac{\rho}{2} \times v^2 \tag{6}$$

The glide ratio is calculated from the distance traveled horizontally divided by the loss of altitude that occurred. It essentially depends on the aircraft type and the so-called aircraft configuration (position of the flaps, landing gear extended or retracted, etc.). In order to model them accurately, the glide ratio for the corresponding aircraft type in the so-called "clean configuration" (landing gear not extended, all flaps, spoilers, etc. retracted to the maximum) must be known, as well as the corresponding factors for each configuration.

The continuously (non-linearly) changing air density mentioned above has other effects in addition to the influence on lift and drag. Aerodynamic lift is the force that counteracts the weight of the aircraft. Since the weight can be assumed to be constant during unpowered flight, a constant lift force is also necessary. From this requirement and the increasing air density during descent, it follows that the relative speed to the surrounding air mass must be greater at high altitudes than close to the ground.

At this point, some terms common in aerodynamics and aviation must be introduced: The true airspeed (TAS), the indicated airspeed (IAS) and the equivalent airspeed (EAS). The TAS is the actual relative speed between the aircraft and the surrounding air mass. If there is no wind, this is identical to the speed over the ground. The IAS, on the other hand, corresponds to the speed which is measured in the aircraft using the so-called pitot tube. This is a forward-facing tube which measures the air pressure due to the speed of movement relative to the air. To determine the IAS, the difference between this dynamic pressure and the static pressure on the aircraft is measured. To do this, the static pressure is taken off at a point on the aircraft that is not affected by the airstream. The speed measured in this way depends on the actual relative speed and the air density. It is usually displayed directly on an instrument in the cockpit. Compared to the TAS, it contains an error, which increases with increasing altitude due to the decreasing air density. The effect of air density on this type of speed measurement now corresponds exactly to the effect that air density has on lift. So in order to fly with a constant lift, you have to fly with a constant IAS. An IAS that is constant over different altitudes in turn causes a TAS that decreases while descending (towards higher air density). At altitudes below 15,000 ft and speeds below Mach 0.3, the EAS is approximately equal to the IAS. In order to simplify the following explanations, we assume $EAS \approx IAS$. Nevertheless, the model we developed can also describe a trajectory based on the exact EAS.

In an accurate model of gliding flights, we need differently defined speeds: the IAS/EAS, which must be kept constant to ensure constant lift, and the TAS, which results from the IAS and the air density (7).

$$TAS = \sqrt{\frac{\rho_0}{\rho(h)}} \times EAS \tag{7}$$

$$with:$$

$$\rho_0 = \text{air density at sea level}$$

$$\rho(h) = \text{air density at altitude h}$$

The TAS is relevant for all geometric aspects of trajectory calculation. This is not just the pure position calculation. The radius r of a coordinated turn depends on the TAS, the gravity of the Earth g and the bank angle of the aircraft.

$$r = \frac{TAS^2}{g \times tan(bank\ angle)} \tag{8}$$

During gliding and the associated decrease in altitude, the TAS as well as the radii of the curves to be flown, constantly change. Some of the trajectory calculation methods based on Dubins paths try to take this into account by taking average values for the values that actually change.

However, due to the non-linear relationships between the variables, this only solves the problems to a very limited extent. The conditions become even more

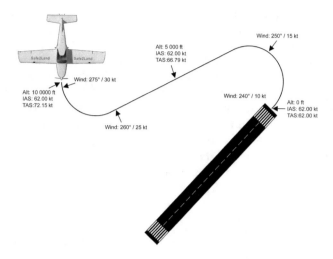

Fig. 8. Example of different IAS, TAS, wind speeds and wind directions over the entire trajectory

complex when you take the wind into account, see Fig. 8. In Dubins-based methods, this is also usually assumed in the form of constant average values for wind speed and direction. In fact, wind speed increases as altitude increases. In addition, the wind direction usually changes with increasing altitude.

Dubins-based methods reach their limits here. In order to model realistic gliding flights, we have developed a completely new approach.

Kinematoid Chains - A New Approach to Transform States Vectors.
The states of the aircraft at the beginning and at the end of the gliding trajectory to be modeled can be described as a vector in a state space Σ.

$$\Sigma = longitude \times latitude \times altitude \times time \times IAS \times TAS... \qquad (9)$$

The parameters longitude, latitude, altitude, time, IAS and TAS are necessary and sufficient for simple models to describe the state. For more precise models, this can be expanded to include additional parameters, such as bank and pitch angle, as well as angular velocities.

The initial state (state when the emergency occurs) must now be transferred to the final state (landing) in such a way that each intermediate state itself as well as the relationships between successive states satisfy certain conditions. Each intermediate state can also be represented as a vector in the state space Σ. Examples of conditions to be met are:

- constant IAS
- maximum bank angle
- constant (or well-defined) decrease in altitude per unit distance traveled

A trajectory can now be modeled as a chain of links, where the chain links represent the states and the joints between them represent the relations of successive states (Fig. 9). The position of the first chain link is now determined by the position of the aircraft, that of the last chain link by the position of the runway. The length of the chain results from the height difference between the aircraft and the runway and the glide ratio E, see (1). In order to determine the positions of all intermediate chain links, the inverse kinematic problem must be solved. This has been well studied in kinematics and computer graphics. The inverse kinematic problem usually has several solutions. In our case, a solution can be found efficiently using the FABRIC algorithm. The basic operation of the FABRIK algorithm is explained in [6]. The application in the context of kinematoid chain is described in [7].

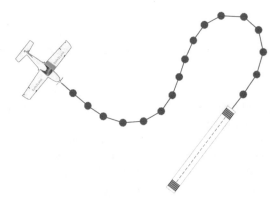

Fig. 9. A trajectory modeled by a kinematoid chain

A trajectory modeled by such a kinematoid chain has a number of advantages over one generated using Dubins path-based methods. By having arbitrarily number of chain links, parameters that change over the trajectory can be modeled with virtually any precision. In this way, the problem of differing IAS and TAS described above can be solved. Both wind speed and wind direction often vary significantly in practice, both vertically and laterally and over time. Conventional methods usually only take wind into account as an average value. The use of kinematoid chains enables precise modeling of complex wind conditions. The prerequisite, of course, is that appropriate data about the prevailing wind speeds and direction are available. As explained in more detail in [7], this method also makes it possible to generate trajectories that avoid obstacles. Here too, methods that compose trajectories from only a small number of elements usually have their limits.

However, the trajectory created using the FABRIK algorithm based on kinematoid chains also has a disadvantage. This method usually results in very long curves with large curve radii, which require a very slight bank angle. This can be unusual for pilots. Particularly in situations of extreme workload, it is desirable not to additionally burden pilots with unusual flight maneuvers. Common flight patterns, for example standard approaches, usually consist of straight sections combined with so-called standard rate turns[1]. Moreover, those maneuvers are also supported by autopilots, which makes it possible to fly an emergency approach autonomously.

To counteract this disadvantage, we have further refined our method. To do this, we first create a trajectory base according to the desired shape. This trajectory is usually not yet practical to fly because it contains, for example, right-angled corners. The aerodynamic conditions described in more detail above have not yet been taken into account. We now develop our kinematoid chains along this trajectory base instead of generating them through the FABRIK algorithm, see Fig. 10. This usually leads to the end point of the generated trajectory being slightly in front of or behind the runway threshold. In an iterative process, the trajectory base is now slightly modified and based on this a kinematoid chain is developed again. The error between the end of the track and the runway threshold is continuously reduced. If the residual error is sufficiently small, the iterative process is aborted. We will then receive an easy-to-fly trajectory that takes precise modeling of the given physical conditions into account.

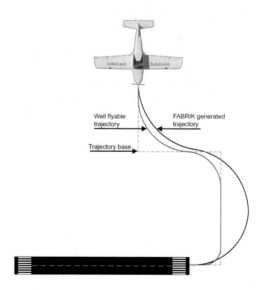

Fig. 10. FABRIK generated vs. well flyable trajectory

---

[1] A standard rate turn is defined as a 3° per second turn, which completes a 360° turn in 2 min.

5   Conclusion

In this keynote we summarized our achievements in the field of engine-out emergency assistance. By the help of our Safe2Land apps pilots can quickly get advice to where and along what route they should glide in case of an engine failure. If no registered airport is within reach we provide an emergency landing field from a database that is automatically generated by AI-methods from public available geodata. In this way, our Safe2Land app can save lives of passengers, crew members, people on the ground and even avoid damage of the airplane.

References

1. U.S. Navy Systems Command: J. Aerodynamics for Naval Aviators, pp. 372–373. CreateSpace Independent Publishing (2015)
2. Thurber, M.: Hands On: Garmin Smart Glide (2022). https://www.ainonline.com/aviation-news/general-aviation/2022-03-01/hands-garmin-smart-glide/. Accessed 18 Jan 2024
3. Davisson, B.: Bundle of energy (2018). https://www.aopa.org/news-and-media/all-news/2018/august/flight-training-magazine/bundle-of-energy/. Accessed 2023
4. Klein, M., Klos, A., Lenhardt, J., Schiffmann, W.: Wind-aware emergency landing assistant based on Dubins curves. In: 2017 Fifth International Symposium on Computing and Networking (CANDAR), pp. 546–550 (2017)
5. Klein, M., Klos, A., Lenhardt, J., Schiffmann, W.: Moving target approach for wind-aware flight path generation. Int. J. Netw. Comput. **8**, 351–366 (2018)
6. Aristidou, A., Lasenby, J.: FABRIK: a fast, iterative solver for the inverse kinematics problem. Graph. Models **73**, 243–260 (2011)
7. Flämig, S., Graefenhan, M., Schiffmann, W.: Modelling of aircraft trajectories for emergency landing using kinematoid chains. CEAS Aeronaut. J. **14**, 679–692 (2023)

The Safety-Related Real-Time Language SafePEARL

Wolfgang A. Halang$^{(\boxtimes)}$

FernUniversität in Hagen, 58084 Hagen, Germany
`wolfgang.halang@fernuni-hagen.de`

Abstract. The programming language PEARL has by far the most distinctive real-time properties, which is why it has excellently proven itself in industrial automation applications. The main features of the language and its support for distributed systems are presented. As the first universally applicable textual programming language, PEARL offers safety-related subsets suitable for each of the four internationally standardised safety integrity levels, as well as special language constructs to formulate safe sequence controls. Parts of PEARL can already be used for specification purposes due to their inherent clarity, unambiguity and direct comprehensibility.

Keywords: Safety · real-time systems · programming language · SafePEARL

1 Introduction

The high-level real-time programming language PEARL (*P*rocess and *E*xperiment *A*utomation *R*ealtime *L*anguage) was developed by automation engineers from around 1969 on. Over the decades, various versions of PEARL were specified and nationally standardised based on the experience gained from several implementations and many industrial applications. The development followed a trend towards more simplicity and less complexity. As there was previously no universally applicable programming language for safety-related real-time systems, it made sense to enable PEARL for this application area. This ultimately led to the current version of PEARL [1] specified in the standard DIN 66253, which provides safety-related subsets for each of the four safety integrity levels standardised internationally in IEC 61508 [4], as well as an extension to formulate safe sequence controls. As will become clear in the course of this chapter, with PEARL a tool is available that is, due to the following properties, extremely well suited to construct autonomously operating cyber-physical systems.

- The language can be learnt by developers in a short time, and can be used effectively to create automation programs thanks to its clear concepts and its unrivalled and application-adequate expressiveness.

- PEARL provides language elements for all tasks occurring in process automation, including input and output of process variables, to limit software costs and risks in process automation projects by being able to completely dispense with the use of assembler languages.
- Despite the fact that process peripherals have a large, manufacturer-dependent variety of implementation forms, PEARL achieves a high degree of portability. This means that automation programs are largely reusable, even if new or different process computer types or makes are used.
- PEARL supports group work in large process automation projects by allowing programs to be broken down into modules that can be developed, compiled and tested independently of each other.
- Programs written in PEARL are self-documenting and easy to maintain, e.g. by syntactically clearly delimitable statement structures, or by data structures that can be built under strict data type binding.
- The algorithmic language core allows simple and clear formulation of numerical calculations of all kinds, such as control algorithms, but also text and bit chain processing.
- Special language tools are used to describe the connections between technical processes and computers as well as the corresponding data and signal traffic.
- The flow of concurrent computing processes is organised using clear, self-documenting real-time language constructs easy for users to understand.
- Clear language constructs allow the programming of distributed computer systems performing automation tasks, including communication between different computers.
- PEARL is the most powerful high-level programming language for automating technical processes and has the most distinct real-time properties.

2   Major Language Properties

Before the individual language features of PEARL are discussed in the following sections, first a coarse impression should be given. For this purpose, the language features of PEARL are divided into five areas:

1. structuring of programs,
2. formulation of algorithmic relationships,
3. input and output,
4. real-time programming and
5. distributed computing.

A PEARL program consists of a *system part* or a *problem part* or both. The system part contains a description of the hardware structure including the connections to the technical process: "system part" actually means "hardware system part". Freely selectable, symbolic names are given to the signals of the process and the hardware modules. The problem part is the actual automation program. It no longer uses special device and signal identifiers, but the symbolic names introduced in the system part.

The purpose of this division into system and problem parts is to collect all special hardware features of a process computer system connected to a technical process in the system part, while the problem part is kept largely independent of this. When transferring a PEARL program to another computer type with a different process periphery, only the system part needs to be changed. In most cases, the problem part, on the other hand, can be transferred almost unchanged, because it only contains symbolic names. This special feature of PEARL achieves extensive portability of automation programs.

The problem parts of PEARL programs can be structured in form of separately compilable program segments, called modules.

The language elements of PEARL for formulating algorithmic relationships and processes correspond to the standard of traditional structured programming languages for technical and scientific applications. In addition to the declaration of new data types, the declaration of new operators is also permitted. Furthermore, in the interest of portability, it is possible to describe the calculation accuracy unambiguously.

In terms of input and output, PEARL contains a standardised concept for process and standard peripherals. It provides the engineer with the input/output instructions required for process applications in a very clear and self-explanatory manner. The following examples illustrate this:

TAKE Pressure **FROM** Manometer;
    (input of the process variable Pressure from the connection Manometer)
SEND UP **TO** Slider;
    (output of the actuating variable UP to the connection Slider)

Of all known programming languages, PEARL offers the most comprehensive, clearest and easiest to understand set of language elements for real-time programming. These include those for

- declaration of computing processes, which are called "tasks" in PEARL,
- control of the transitions between task states,
- one-time or cyclically repeated scheduling of tasks depending on time conditions or the arrival of interrupt signals and
- synchronisation of tasks.

The following examples illustrate the clarity of the instructions for describing temporal sequences:

AFTER 5 **SEC ALL** 7 **SEC DURING** 106 **MIN ACTIVATE** Relay **PRIORITY** 5;
(The task Relay is started periodically after 5 seconds every 7 seconds for 106 minutes with priority 5, i.e. it is transferred to the state "ready".)
AT 12:00:00 **ALL** 1 **MIN UNTIL** 12:59:00 **ACTIVATE** Measurement;
(The task Measurement is started every minute from 1200 hrs. until 1259 hrs.)

A separate descriptive part, the so-called *architecture description*, serves as a program component of PEARL modules for programming distributed computer systems. It describes the involved computers called substations, the connections to the peripherals and the technical process as well as the allocation of program parts to individual computers. This software configuration also includes information on dynamic reconfiguration, i.e. the automatic relocation of program parts of a faulty substation to achieve fault tolerance.

In the following sections, the language properties of PEARL are explained in a simplified and abbreviated manner in order to illustrate the essence of real-time programming.

2.1 Program Structure

A PEARL program can be divided into a number of independently compilable units. Together with routines of an operating system and a runtime system, these modules form an executable PEARL program system. The system part describes the device configuration of a process control computer and its connections to the technical process. The problem part contains the actual automation programs. They can be described independently of the computer using the names defined in the system part, which must be defined in the problem part before they are used. PEARL distinguishes between two types of such definitions:

1. A new program variable to be introduced is defined with a *declaration*. This is done using the keyword **DECLARE**, abbreviated to **DCL**. If a program variable is to be known in several modules (definition at module level), then this must be indicated using the attribute **GLOBAL**.
2. A *specification* is used to define a program variable already known. The keyword **SPECIFY**, abbreviated to **SPC**, is used for this purpose. A specification is required, for instance, if a problem part refers to data to which names are already assigned in the system or problem part of another module. In this case, one also speaks of the import of program variables that are exported by other modules.

A character set consisting of the 52 upper and lower case letters, the ten digits and some special characters is permitted for programming in PEARL. Names of program variables (identifiers) are formed from letters and numbers, whereby a name's first character must be a letter. Comments can be inserted between language elements in the program text, i.e. explanations to understand a program better. They are enclosed in the two end symbols /* and */ or introduced as line comments by !.

Statements in a PEARL program represent self-contained operations, and are separated from each other by semicolons. This makes it possible to write an instruction over several lines as well as several ones in a single line, e.g.

DCL (XYZ, UVW, ABC) **FLOAT**;
XYZ:=1.0; UVW:=−2.0; ABC:=3.0;

Table 1. Examples of representing constants in PEARL

Data type	Constant consists of	Example
FIXED	one or several decimal digits	1023
FLOAT	floating point number in common notation floating point number in exponential notation	0.1023 10.23E−2
BIT (n)	one or several dual digits enclosed in apostrophes, followed by the designation **B1**; one or several octal digits enclosed in apostrophes, followed by the designation **B3**; one or several sedecimal digits enclosed in apostrophes, followed by the designation **B4**;	'1111111'**B1** '1777'**B3** 'A174'**B4**
CHAR (n)	one or several characters of the entire character set enclosed in apostrophes	'Text'
CLOCK	time in hours, minutes and seconds with ":" as delimiter	18:36:17
DURATION	duration in hours, minutes and seconds with **HRS**, **MIN** and **SEC** as delimiters	2 **HRS** 25 **MIN** 31 **SEC**

Within a problem part, for structuring purposes tasks and procedures are defined, which can be further subdivided into blocks to be introduced by the keyword **BEGIN** and terminated by **END**.

For modules, tasks, procedures and blocks apply *visibility rules* with regard to the data objects occurring in them, which are defined locally within a task, a procedure or within a block, and are not known in other parts of the program. On the other hand, these data objects are known in any subblocks.

Data objects can be individual data of a basic data type. They can, however, also be composite ones such as records and matrices. Data being changed during the execution of a program are referred to as variables. Data that cannot change during program execution are called constants. They must be identified by the keyword **INVARIANT**, abbreviated to **INV**. PEARL has the following data types for variables and constants; examples of representing constants of these data types are shown in Table 1:

FIXED	interger number
FLOAT	floating point number
BIT (n)	binary pattern with n bits (bit chain)
CHAR (n)	character sequence with n characters (character string)
CLOCK	(point in) time (instant)
DURATION	duration (of time)

Variables and constants must be declared or specified, respectively, before they are used in tasks or procedures. For this, names and data types of the data objects must be indicated. Initial values of variables are defined with the keyword **INITIAL**, abbreviated to **INIT**, e.g.

Table 2. Examples for simple arithmetical and Boolean operators in PEARL

Arithmetic		Comparison		Boolean operators	
Op.	Operation	Op.	Operation	Op.	Operation
+	addition	**LE**	$\leq$ smaller or equal	**NOT**	not (negation)
-	subtraction	**GE**	$\geq$ greater or equal	**AND**	and (conjunction)
/	division	**LT**	$<$ smaller than	**OR**	or (disjunction)
*	multiplication	**GT**	$>$ greater than	**EXOR**	exclusive or (antivalence)
**	exponentiation	**EQ**	$=$ equal		
//	integer division	**NE**	$\neq$ unequal		

Declaration of the floating point variables A1, A2, A3, A4 with the initial values A1=0.1, A2=0.2, A3=0.3, A4=0.4:
DECLARE (A1, A2, A3, A4) **FLOAT INIT** (0.1,0.2,0.3,0.4);
Assignment of the name PI to the constant 3.14159:
DCL PI **INV FLOAT INIT** (3.14159);

2.2   Algorithmics

As language constructs for programming algorithms, which are not specific to process automation, PEARL provides various types of statements, similar to other programming languages, viz.

– assignments,
– control flow statements and
– procedure calls.

With an assignment, any values of specific types can be assigned to variables during program execution. Both the equal sign "=" and the sign ":=" are permitted as assignment symbols. To the left of the assignment symbol stands the variable to which a value is to be assigned, to the right an expression consisting of one or more operands linked by arithmetic operators. The operators can be used to perform arithmetic and logical operations. Table 2 gives examples of operators in PEARL.

Normally, a program's instructions are executed in the order in which they follow each other. Control flow statements are used to specify a different sequence. There are two types of control flow statements in PEARL:

1. statements to branch the program flow and
2. statements for repetitive execution of statements in loops.

To branch the program flow PEARL provides three options:

1. choice between two alternative paths of control flow,
2. case selection among an arbitrary number of alternatives and

3. jump statement.

With the first one

IF expression of type **BIT(1)**;
   THEN statement(s)1
   [**ELSE** statement(s)2]
FIN;

the **THEN** branch is taken if the Boolean expression yields the result true ('1'**B**). If the condition is false (result '0'**B**), the **ELSE** branch is executed, in case there is one. With the case selection

CASE expression of type **FIXED**;
     ALT statement(s)1
     ALT statement(s)2
     ...
     [**ALT** statement(s)n]
     [**OUT** statement(s)]
FIN;

the first, second, third etc. program branch is run through if the expression behind the keyword **CASE** assumes the value 1, 2, 3, etc., respectively. If this value is not between 1 and the number of specified alternatives then, if applicable, the statement sequence specified in the **OUT** clause is executed. The jump instruction **GOTO** can be used within the block structure to branch from any point in the program to another instruction that is preceded by a label indicating a jump target. Jump labels are formed like variable names and are declared by placing them in front of an instruction, separated by a colon. The jump target must either be in the same block or in an enclosing block. Jumps into blocks are not permitted.

Despite this restriction, the use of jump instructions is potentially dangerous, because it may lead to the creation of complex and error-prone programs. Jump instructions are, therefore, not permitted in structured programming. Any program containing jump instructions can be converted into a structured form without them—however at the cost of increased computing time and memory requirements.

Repetition or loop statements have in PEARL the general form

[**FOR**   loop-variable-name]
[**FROM**   expression of type **FIXED** (initial value)]
[**BY**   expression of type **FIXED** (increment)]
[**TO**   expression of type **FIXED** (final value)]
[**WHILE** expression of type **BIT(1)**]
REPEAT [loop-body-statement(s)]
END;

and are used for the repeated execution of instructions or instruction sequences. The loop body written between the keywords **REPEAT** and **END** is executed again and again as long as the expression of the **WHILE** clause has the value '1'**B** and (in terms of logical conjunction) the loop variable has not exceeded the final value, yet.

Procedures are used in the formulation of algorithmic relationships so that instruction sequences occurring several times in a program do not have to be written down several times. Once an algorithm is programmed, it can be reused in other program parts. Procedures are, therefore, also used to structure programs. Proper and function procedures can be created in PEARL. Proper procedures cause certain sequences of instructions to be executed, possibly depending on certain parameters. A proper procedure is defined as follows:

Name: **PROCEDURE** (List of parameters and respective data types);
       definitions;
       [statement(s)]
       END

Example of defining a proper procedure called TBA:

TBA: **PROCEDURE** (L **BIT(1)**, A **FIXED IDENTICAL**, B **FIXED**);
     IF L **EQ** '1'**B1 THEN** A:=A+B; **ELSE** A:=A*B; **FIN**;
     END;

The procedure is invoked with a **CALL** statement in which other, so-called actual parameters are given instead of the declared ones. In this example, the procedure is called by the statement:

CALL TBA (F,C,D); or just TBA (F,C,D);

The actual parameters F and D are passed as values (call by value), i.e. there are two new memory cells in the procedure into which the values of F and D are stored. Changes to F and D in the procedure do not affect the calling program. The attribute **IDENTICAL** of the parameter A causes the address of its memory location to be passed to the procedure (call by reference). Then, the procedure and the calling program access the same memory cell. A change to A in the procedure is, therefore, passed on to the calling program. The actual parameters F, C and D thus introduced must be declared in the calling program with the types **BIT(1)** (for F) or **FIXED** (for C and D). The **CALL** statement then causes the following program segment to be executed:

IF F **EQ** '1'**B1 THEN** C:=C+D; **ELSE** C:=C*D; **FIN**;

Function procedures calculate a (function) value from one or several input parameters. A function procedure's general form is:

Name: **PROCEDURE** (List of input parameters and respective data types)
       RETURNS (data type of the function value);
       definitions;
       [statement(s)]
       RETURN (expression);
       END

A function procedure calculating an integer number's factorial may read as:

```
FACTORIAL:  PROCEDURE(N FIXED) RETURNS(FIXED);
            DECLARE NF FIXED INIT(1);
            IF N EQ 0
              THEN RETURN (NF);
              ELSE
                FOR I FROM 1 BY 1 TO N
                  REPEAT NF:=NF*I;
                END;
            FIN;
            RETURN (NF);
            END;
```

This procedure is invoked within a program as, e.g.

K:=FACTORIAL (M);

where M must be declared in the invoking program as an integer variable. By this call M! is computed and the result is assigned to the variable K.

2.3   Process Data Input and Output

The input and output of process data is of great importance in process automation. Actual values of technical processes are recorded by sensors, important information is output to the operating personnel in control rooms, from there instructions by the operating personnel are received, and the technical processes are influenced by means of actuators. The data exchange with files on mass storage devices that takes place within process computers is also counted as input/output. The input/output capabilities are the links between a process computer and the program executed on it on the one hand, and the technical process with its sensors and actuators as well as the control room or the devices connected to it on the other. However, as input/output is device-dependent, i.e. differs from one process computer to the other, PEARL provides a comprehensive concept minimising the effects of these differences.

The central means of describing input/output is the so-called *data station*, also known as **DATION** for short. A data station is assigned to any sensor/actuator and any standard peripheral device. In problem parts, with input/output instructions data objects are only read from data stations or send out to them. Real devices or their driver programs are assigned to data stations in system parts. This ensures that the information dependent on the respective hardware of a process computer, i.e. the one which must be changed when porting a program to another computer, can practically all be found in the program's system part, and that its problem part can be adopted (largely) unchanged. The exact language constructs for describing the connections between the devices of a process computer, the assignment of the device connections to the data stations, and the direction of the information flow at the connection points are now explained. Device connections are described as:

device-identifier symbol for the transmission direction device identifier; The symbol indicating transmission directions are:

A -> B   transmission from A to B,
A <- B   transmission from B to A,
A <-> B   transmission in both direction.

If several other devices are connected to a device, in addition to its name also the number of the connection, marked with an asterisk and followed by the number of the connection, is given. The connection between a device control unit and a central unit (CU) is described as:

CU*0 <-> device-control-unit;

The device control unit is connected to connector 0 of the central unit. A digital output is connected to the device control unit accordingly:

device-control-unit*6 -> DigOut;

The specification of data flow directions allows the translator to recognise contradictions in the use of connections. In this example, the use of DigOut for input would be recognised as an error.

Syntactically, a data station is assigned to a physically existing device in the following manner, which describes the connection between the "PEARL world" with its (abstract) data stations and the real field devices:

data-station :   device-identifier;

Freely definable symbolic names can be selected as identifiers for data stations. If several data stations can be connected to one device, in addition to the identifier the number of the connecting terminal and the number of bits that the data station comprises are specified, e.g.:

Scale: DigOut*8,1;
Weight: DigIn*2,10; /* 10 = number of connected bits */

Interrupt signals report external events that are usually of particular importance. Their description is similar to the one just explained, e.g.:

Aready: IrtpIn*2;
DEready: IrtpIn*3;

Indications of internal events, i.e. events that occur within a process computer system, are referred to as *signals*. For example, the message "division by zero" is a signal that may be generated internally during a program run. A signal is described in a very similar way to an interrupt:

signal-name :   identifier-of-internal-event;

e.g.:

InputError: ErrorType(4);

Once data stations with corresponding identifiers are introduced in a system part, they can be used within the associated problem part. First, however, in accordance with the visibility rules explained above, they must be made known in the problem part, i.e. specified (**SPECIFY**). In addition to the system-defined data stations for the standard periphery devices (printer, mass storage, operating console, etc.) or the process periphery (sensors, actuators, etc.), user-defined data stations can be introduced in the problem part. These are data stations created on system-defined data stations, e.g. a (user-defined) file for the temporary storage of data on a (system-defined) magnetic disk. As user-defined data stations are introduced for the first time in the problem part, it is not sufficient to specify them, i.e. to make them known within the problem part. Instead, declarations (**DECLARE**) of these data stations are required. The case just mentioned that a file is to be created on a magnetic disk is described in PEARL as:

```
MODULE;
  SYSTEM;
    DISK: PSP31;
    . . .
  PROBLEM;
    SPECIFY DISK DATION INOUT ALL;
    DECLARE FILE DATION INOUT FLOAT DIRECT CREATED(DISK);
```

Apart from making known or introducing a data station, its properties are thus described in the problem part. The magnetic disk is made known as a data station with the designation DISK in the problem part, whereby both reading and writing is possible (**INOUT**). From the outset, the data types used are not restricted (**ALL**). The most important options for describing data stations are summarised in Table 3.

Table 3. The most important properties of PEARL data stations

Property	Explanation
transmission direction	**IN** for input devices **OUT** for output devices **INOUT** for dialogue devices, peripheral storage etc.
data representation	the data types a data station can accommodate, e.g. **ALPHIC** alphanumerical characters, **BASIC** process data, **ALL** all possible types
type of access	kind of positioning, e.g. **FORWARD** for a sequential file or **DIRECT** to position a cursor
control channel	**CONTROL(ALL)** is indicated for explicit positioning or to format output data

A basic distinction is made between three types of data stations, to which different input/output instructions are assigned. The kind of data type is indicated in the specification or declaration in the problem part:
ALPHIC for input and output of alphanumeric characters. Typical data stations of this type are display devices and printers. The associated input/output instructions **GET** and **PUT** cause a conversion between the character set used in the data station and the computer's internal format. Example:

> **SPECIFY** Printer **DATION OUT ALPHIC CONTROL(ALL);**
> **PUT** 'The temperature is', Temperature **TO** Printer;

BASIC for the input/output of binary data from and to the process periphery, e.g. for reading a switch status. The associated input/output instructions **TAKE** and **SEND** comprise a conversion between the representation as binary sequences and the computer-internal one. Example:

> **SPECIFY** Temperature-sensor **DATION IN BASIC;**
> **TAKE** raw-temperature **FROM** Temperature-sensor;

Internal (**FIXED, FLOAT, BIT, ALL** etc.) for computer-internal data exchange, i.e. there is no conversion of the data format during input/output. Here, the type of data to be transferred is indicated in the specification or declaration of the data station in the problem part. The corresponding input/output instructions are **READ** and **WRITE**. Example:

> **SPECIFY** Disk **DATION INOUT ALL;**
> **DECLARE** File **DATION INOUT FLOAT CREATED** (Disk);
> **WRITE** Temperature **TO** File;

To simplify matters, the three types of data stations can be associated with input/output from and to control rooms (**ALPHIC**), from and to technical processes (**BASIC**), and data exchange with mass storage devices (internal). In principle, the general structure of input/output instructions can be explained with the two forms

> **PUT** expression-list **TO** data-station-identifier **BY** format-list;
> **GET** object-list **FROM** data-station-identifier **BY** format-list;

wherein, depending on the type of data station, **GET** or **PUT** are to be replaced by **TAKE** or **READ** or **SEND** or **WRITE**, respectively.

For input/output, there are often special requirements to make data available to the operating personnel in formatted form such as tables. For this reason, there are numerous options for the format list in the input/output instructions. Some important ones are briefly outlined using an example:

> **SPECIFY** Laser-printer **DATION OUT ALPHIC;**
> **PUT** 'Number:', Number, 'Cost:', Cost **TO** Laser-printer
>   **BY** X(2), A(7), X(2), F(4), X(2), A(7), X(2), F(6,2);

Table 4. Format elements with examples

Data type	Format element	Significance of format element	Example	
			Data	Format
FLOAT or FIXED	**F(w,d)** or **F(w)**	I/O of an integer with w digits incl. sign, decimal point and d digits behind it	427.35 −30	**F(6,2)F(3)**
FLOAT	**E(w,d)**	I/O of a real number with w digits in exponential notation with d mantissa digits to the right of the decimal point	−4.27E−10	**E(9,2)**
BIT	**B1(w)**	I/O of a bit chain as binary number with w digits	1001001	**B1(7)**
BIT	**B3(w)**	I/O of a bit chain as octal number with w digits	765	**B3(3)**
CHAR	**A(w)**	I/O of a string with w characters	new set points (in degrees C)	**A(26)**
CLOCK	**T(w)**	I/O of a clock time, in total with w characters	10:40:47.63	**T(11)**
DURATION	**D(w)**	I/O of a duration, in total with w characters	20 **HRS** 7 **MIN** 21 **SEC**	**D(15)**
control elements	**X(w)** **SKIP** **PAGE**	output of w spaces begin of a new line begin of a new page		

The texts between the apostrophes, in this case "Number:" and "Cost:", are printed as indicated by the corresponding alphanumeric control formats (**A(7)**), namely as 7 characters long strings each (including any added spaces). The statement **X** generates an indicated number of spaces for the **FORWARD** access type, and the **F** control statements determine for **FIXED** or **FLOAT** quantities the total number of available and the number of possible decimal places; e.g. **F(6,2)** describes the output of a **FLOAT** number 6 digits long with 2 of them behind the decimal point. Hence, the above-mentioned output statement generates the following printout on the laser printer:

    Number: 20 Cost: 420.80

Table 4 shows a compilation of all different format instructions for the input/output of various data types with examples.

2.4   Real-Time Programming

PEARL's language constructs for real-time programming can be divided into five groups:

1. declaration and specification of concurrent computing processes (tasks),

2. controlling the transitions between task states,
3. scheduling of tasks,
4. synchronisation of tasks and
5. interrupt statements.

Tasks are declared statically in the form

Task-name: **TASK;**
                task-internal definitions;
                statements;
                END;

The transitions between task states are controlled by the following statements as depicted in Fig. 1:

Activation ACTIVATE Task-name [**PRIORITY** Priority-number];

The mentioned task is transferred into the state "ready", i.e. declared ready to be executed.

Termination TERMINATE [Task-name];

The designated task or the task executing the instruction is transferred from the "running" or "ready" state into the state "dormant".

Suspension SUSPEND [Task-name];

The designated task or the task executing the instruction is transferred from the state "ready" into the state "blocked".

Continuation CONTINUE [Task-name];

The designated task or the task executing the instruction is transferred from the state "blocked" into the state "ready", i.e. may be executed again.

Deferment Event **RESUME;**

The task executing the instruction is transferred to the state "blocked" and transferred back to state "ready", i.e. is continued, after the specified event occurs, e.g. when an interrupt signal arrives or a time period expires.

De-scheduling PREVENT [Task-name];

The schedules for future activations of the designated task or the task executing the statement are deleted, i.e. the task's start conditions become ineffective. A running task is not cancelled by this command.

The transition between the task states "ready" and "running" shown as a dashed line in Fig. 1 cannot be brought about by PEARL statements, but only by the operating system, as only the latter can decide which of the computing processes in the "ready" state has the highest priority and is, therefore, the first one to be transferred to the "running" state. This is not known when a PEARL program is written, so it cannot be indicated in the program.

Declaration of a task

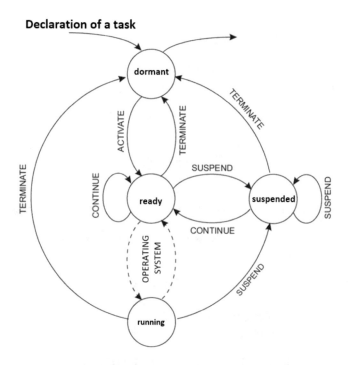

Fig. 1. A task's four basic states and the PEARL statements controlling the state transitions

Constants and variables of the data types time and duration are defined in PEARL for the scheduling of tasks. A time constant denotes a fixed point in time in the form:

hours : minutes : seconds
Example: 1:36:25 means 1 o'clock 36 min 25 sec

The value of a clock variable is a point in time (moment). The declaration of such a variable reads

DCL Variable-name **CLOCK;**
Example: **DCL** Time **CLOCK;**

A time duration constant denotes a fixed duration of time. It can consist of hours, minutes and seconds and has the form

zh **HRS** zm **MIN** zs **SEC**
Example: 2 **HRS** 25 **MIN** 10 **SEC** means 2 hours 25 minutes 10 seconds

The value of a duration variable is a time duration. The declaration of such a variable reads

DCL Variable-name **DURATION;**
Example: **DCL** Waiting-time **DURATION;**

The periodic activation of tasks at certain points in time is scheduled with the following clause:

AT clock-expression /* moment for which the first execution is requested */
ALL duration-expression /* period */
UNTIL clock-expression /* moment for which the last execution is requested */
Example:
AT 12:00:00 **ALL** 1 **MIN UNTIL** 12:59:00 **ACTIVATE** Measurement;

The periodic activation of tasks in certain time intervals is scheduled with the following clause:

AFTER duration-expression /* duration until the first execution is requested */
ALL duration-expression /* period */
DURING duration-expression /* duration until the last execution requested */
Example: **AFTER** 5 **SEC ALL** 7 **SEC DURING** 106 **MIN ACTIVATE** Relay **PRIORITY** 5;

A task, which is to be executed depending on the occurrence of an interrupt signal and periodically afterwards, is scheduled with the following clause:

WHEN interrupt-signal-name /* and then optionally */
ALL duration-expression
UNTIL clock-expression /* or */ **DURING** duration-expression
Example:
WHEN IRPT **ALL** 1 SEC **UNTIL** 5 **MIN ACTIVATE** Measurement;

Table 5 shows examples of applying the keywords for task control.
Semaphores can be employed to synchronise tasks. With the operation

REQUEST Sema-variable;

the value of the semaphore variable is decreased by 1 if the result is not negative. If the result is negative, however, the operation is postponed and the execution of the task containing the instruction is put on hold until the value of the semaphore variable becomes positive again. In this case, the simple state diagram shown in Fig. 1 is no longer sufficient to describe the task states. Instead, an additional state "blocked by synchronisation operation" must be introduced there. The operation

RELEASE Sema-variable;

Table 5. Examples of applying the keywords for task scheduling and control

Statement	Effect
ACTIVATE A;	Task A is transferred to the state "ready".
AFTER 5 **SEC** **ALL** 7 **SEC DURING** 106 **SEC** **ACTIVATE** B;	Task B is activated for the first time 5 seconds after executing the activation statement, and from then on cyclically for 106 seconds with period 7 seconds.
AT 10:00:00 **ALL** 30 **MIN** **ACTIVATE** C **PRIORITY** 1;	Beginning at 10 o'clock task C is activated every 30 minutes with priority 1.
WHEN ALARM **ACTIVATE** D **PRIORITY** 3;	Upon occurrence of the interrupt signal ALARM task D is activated with priority 3.
WHEN WARN **ALL** 1 **MIN UNTIL** 11:00:00 **ACTIVATE** E **PRIORITY** 3;	Upon occurrence of the interrupt signal WARN task E is activated until 11 o'clock with period 1 minutes and priority 3.
F: **TASK**; . . **SUSPEND**;	 . . Task F is transferred to the state "blocked".
SUSPEND G;	Task G is transferred to the state "blocked".

increases the value of the semaphore variable by 1.

In a technical process, it may be necessary to suppress the effect of interrupt requests, i.e. the start conditions for tasking statements relating to them should be invalid when an interrupt occurs. For this purpose, one uses the instruction

DISABLE Interrupt-signal-name;

which is cancelled again by

ENABLE Interrupt-signal-name;

An incoming interrupt signal with the mentioned name is then heeded again.

3 PEARL for Distributed Systems

PEARL provides suitable language tools for programming distributed real-time systems—features completely missing in all other common programming languages. On the one hand, these are language constructs to describe the hardware set-up of distributed systems, i.e. of the

- process computer units involved, called stations,
- physical connections between the computers and
- connections to the periphery and the technical process.

There are also language constructs to describe the software properties

- distribution of program modules to stations,

- communication connections between program modules (logical connections),
- types of communication flow (transmission protocols) and
- reaction to faulty states of distributed systems (failure of stations or connections).

In the event of a fault, the redundancy present in a distributed system can be utilised in the sense of functional redundancy. This requires the ability for *dynamic reconfiguration*, i.e. the automatic relocation of program parts from a faulty automation unit to a functional one. Such reconfiguration ought to enable a computer system to react to faulty states *during operation*—hence the term dynamic reconfiguration—with the aim to maintain the functionality of the automation system at all times.

With the architecture description, a higher-order description list is available in PEARL as a program component to describe distributed systems. Distributed programs are structured by combining their modules into module groups called *Collections*. Depending on certain operating states such as normal operation or failure of one or more computers or connections, collections are distributed to the individual stations and run there. The collections are the reconfigurable units for dynamic reconfiguration. A PEARL program thus consists of a higher-order architecture description and several collections which, in turn, consist of one or more modules.

An architecture description is divided into four main components and is structured as:

ARCHITECTURE;

Station part:	description of the distributed system's stations
Network part:	description of the physical connection paths
System part:	description of access to the periphery
Configuration part:	description of the software configuration

ARCHEND;

The instructions in the architecture description are required to control and parameterise the system programs (compiler, linker, loader) during program generation and for the runtime system. The architecture description is used exclusively for this purpose and is not converted into machine code nor an executable program. The module groups, for example, are translated for different target computers according to this description.

In the station part, all stations of a distributed system are described by specifying their names, attributes and their possible operating states:

STATIONS;
    NAME: User-name;
    PROCTYP: Processor-type;
    WORKSTORE: SIZE Main-storage-size;

    . . .

    STATEID: Assignment of identifiers to operating states;
STAEND;

The name attribute user name or station identifier is employed to identify a station within a communication structure. A network can be made up of different computers (inhomogeneous system). Additional information about the properties of the individual target computers, such as processor type and memory size, is, therefore, necessary for the compiler. Specifying the stations' operating states is the basis for possible monitoring mechanisms in fault-tolerant multi-computer systems. By querying the current operating states, it is possible to react to hardware or software errors by dynamically reconfiguring the software.

Within a distributed computer system, communication can take place via any arbitrarily complex connection structure such as bus, ring, star or network structures or combinations thereof. The lines may be suitable for unidirectional or bidirectional data exchange. The physical connection structure of a communication system is described in the *network part*. A distributed network operating system uses this route information to forward message packets to their destination stations by means of destination addresses specified in the packet headers. Similar to the description of station-local connections between devices of a station in the system part, cross-station connections are described in the network part. A physical connection or "line" is defined by specifying its end points with the corresponding station identifiers, a transmission direction symbol (<-, ->, <->) and its name. Figure 2 shows an example of the description of a physical connection structure.

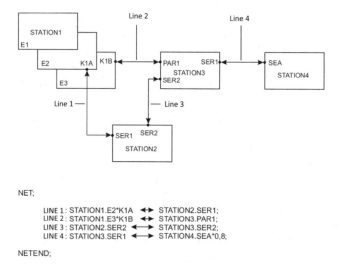

Fig. 2. Description of a physical connection structure between the stations 1 through 4; a connection's end point can be described by a station identification (e.g. Station1), a station-internal identification of a plug-in unit (e.g. E2) or by indicating a connector (e.g. K1A).

In the system parts of PEARL programs running on single-processor systems, periphery access is defined by assigning user names to the peripheral devices. The validity areas of the user names declared in the system parts are the entire programs, every one consisting of one or more modules. In order to be able to reuse such programs in distributed systems as well, it is possible to define single-processor PEARL programs as collections. The validity area of the system subobjects then comprises at most one collection.

In fault-tolerant multi-computer systems with dynamic reconfiguration of the software, however, the device description of a program part's system part depends on the station to which this program part is loaded. Therefore, the concept that a system part represents a component of a collection does not make sense for reconfigurable systems. For this reason, as a second alternative for distributed systems, system parts can be removed from modules and placed in the system part of the higher-order architecture description. In this case, only one system part is required per station.

The software configuration, i.e. the distribution of collections to individual stations, and dynamic reconfigurations are defined in the *configuration part*. An example of a configuration part is shown in Fig. 3. There, several computers are used for the automation of a production line with several production steps (milling, drilling etc.). In the example, only the configuration for drilling automation and the overall operation monitoring are shown.

A configuration part has an *initial part* and may contain a *reconfiguration part*. The configurable units (collections) and their interfaces (ports) are defined in the initial part. Furthermore, the configuration for normal, undisturbed operation is described using **LOAD** and **CONNECT** statements. The reconfiguration part uses **DISCONNECT** statements to dismount connections in the course of dynamic reconfigurations depending on operating states (**STATE**), **REMOVE** and **LOAD** statements to relocate collections, and **CONNECT** statements to establish new logical connections.

Module groups (collections) communicate exclusively by exchanging messages on the basis of the *port concept*. In problem parts, messages exchanged between collections are addressed solely by specifying ports. The ports thus represent the interfaces between collections. A distinction is made between input (receive) and output (send) ports. By connecting an output with n inputs, $(1 \rightarrow n;$ 1 sender, n receivers) or, conversely, $(n \rightarrow 1)$ communication structures can also be set up. A collection may contain several ports.

The logical connections between collections via input and output ports are described using **CONNECT** statements in the configuration part. Different **CONNECT** statements may be defined for different configurations in the reconfiguration part. In this way, collections can be used in changing environments, see Fig. 3 for reconfiguration after failure of Station2. Also in **CONNECT** statements, logical connections can be mapped to physical connections by specifying lines defined in the network part. The keyword **VIA** is used to select predefined lines, the keyword **PREFER** to select preferred lines, or no specifications may be made. In the latter case, the network operating system must determine functioning and free lines itself.

```
CONFIGURATION;
/*Initial part*/
  COLLECTION DrillAutomation
    MODULES FeedControl, CarriageControl, RotationSpeedControl
    PORTS   PDrill1, PDrill2, PDrill3;

  COLLECTION OperationMonitoring
    MODULES DrillDataEvaluation, Statistics, DataLogging
    PORTS   PMon1, PMon2, PMon3, PMon4;

  LOAD DrillAutomation TO Station1;
  LOAD OperationMonitoring TO Station2;

  CONNECT DrillAutomation.PDrill1 ->
              OperationMonitoring.PMon1;
  CONNECT DrillAutomation.PDrill2 ->
              OperationMonitoring.PMon2 PREFER Line1;

/*Reconfiguration part*/
  STATE (Station2.Fail AND NOT Station3.Fail)
    BEGIN

      DISCONNECT DrillAutomation.PDrill1 ->
                    OperationMonitoring.PMon1;
      DISCONNECT DrillAutomation.PDrill2 ->
                    OperationMonitoring.PMon2;

      REMOVE OperationMonitoring FROM Station2;

      LOAD OperationMonitoring TO Station3;

      CONNECT DrillAutomation.PDrill1 ->
                  OperationMonitoring.PMon1 VIA Line2;
      CONNECT DrillAutomation.PDrill2 ->
                  OperationMonitoring.PMon2 VIA Line2;

    END;
CONFEND;
```

Fig. 3. Example of a configuration part with an initial and a reconfiguration part employing the connection structure of Fig. 4

Before using a **PORT** object in the problem part of an automation program, this port must be specified in the corresponding module. The specification of ports and the send (**TRANSMIT**) and receive (**RECEIVE**) operations within a task are shown in Fig. 4 using an example. When specifying a port, the following attributes can be assigned to it:

- transmission direction **IN** or **OUT**,
- message type,
- protocol type,
- possible information about the buffer area of the receive port and
- exception handling.

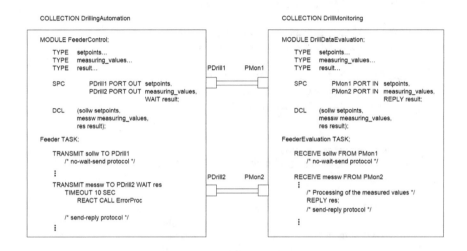

Fig. 4. Examples of specifying ports and of transmit and receive operations

If the task `Feed` wants to send a message, e.g. nominal or measured values, to the task `FeedEvaluation`, it contacts a send port `PDrill1` or `PDrill2`. According to Fig. 3, the send port is connected to a receive port and transmits the message to it. The task `FeedEvaluation` uses a **RECEIVE** operation to retrieve the message at the receive port `PMon1` or `PMon2`, respectively. If the receive buffer does not contain a message upon entering into a **RECEIVE** instruction, the receiving task is postponed until a message has arrived. A **TIMEOUT** clause can be used to limit the blocking of a task to a certain period of time. Once this duration has expired, it is branched to the **REACT** statement. To handle transmission errors, signals can be assigned to a port that allow for suitable error reactions. There is a number of predefined signals, such as the indication of a full buffer at the receive port.

Three protocol variants are available for handling communication:

1. Asynchronous communication with the "no-wait-send" protocol. With a send operation, the message is passed to the network operating system, which carries out the transmission and, if necessary, buffering on the receiving side. An example of this is the communication via the ports `PDrill1` and `PMon1` in Fig. 4.

2. Synchronous communication with the "blocking-send" protocol. The transmission data of a sending task are transferred to the network operating system, and this reports readiness to send at the receive port. When the receiving task reaches a **RECEIVE** instruction, readiness to receive is acknowledged and the data can be transmitted or a **TRANSMIT** instruction can be executed.
3. Synchronous communication with the "send-reply" protocol. Here, sender and receiver have a contractual relationship with each other. The sender waits for its message to be processed, e.g. the measured values in Fig. 4, and receives a result message back. This protocol must not be processed via a $(1 \to n)$ connection.

The types of synchronous communication are identified by the keywords **WAIT** and **REPLY**. If a type for retransmission data is specified, the connection is handled by the "send-reply" protocol, cp. communication via the ports PDrill2 and PMon2 in Fig. 4. With the "blocking-send" protocol, however, the retransmission is omitted. Retransmission (acknowledgement) takes place via the same connection, i.e. from an **IN** port to an **OUT** port. Time restriction mechanisms may be applied to both **TRANSMIT** and **RECEIVE** operations for synchronous communication.

The transmission protocols also allow to realise $(1 \to n)$ or $(n \to 1)$ communication structures with the exception of the "send-reply" protocol. In an $(n \to 1)$ communication structure, messages from different senders are waited for simultaneously. If several send ports have already signalled readiness to send upon entry into the **RECEIVE** statement, or if messages are available, one of the senders is selected using a non-deterministic procedure. If there are no transmission requests, however, the system waits for one of the n ports to report readiness to send. Since all ports involved must have the same message type, alternative waiting is restricted to messages of the same type. In order to wait alternatively for messages of different types, n ports must also be used on the receiving side and a selected-**RECEIVE** instruction must be used:

```
RECEIVE
  SELECT nominal-value FROM PMon1
  OR measured-value FROM PMon2
  /* Processing of the measured values */
  REPLY result;
```

4 Safety-Related PEARL

As there was previously no universally applicable textual programming language specifically for safety-related real-time systems, but PEARL had already proven its excellent feasibility to formulate industrial automation applications, a suitable subset of PEARL for each of the four internationally standardised safety integrity levels, and an extension to describe safe sequence controls were defined, and standardised in 2018 [1]. These sublanguages are presented in this section. Their syntactic form follows the tradition of previous PEARL versions, i.e. the formalisation is extremely low, so that even the new language constructs can be read immediately and interpreted unambiguously by engineers without prior training.

4.1   Verification-Oriented Language Subsets

In order to be able to prove the correctness of programs with maximum trustworthiness and minimum effort, it is necessary to use programming concepts that support the verification process as much as possible. Verification and, therefore, the assurance that a program is error-free, is in fact the basic condition for issuing a safety license. Conditions and goals to be fulfilled by safety-related automation systems can only be achieved if *simplicity* is chosen as fundamental design principle and—usually artificial—complexity is combated, because simple systems are easy to understand and their behaviour is easy to comprehend. This corresponds to the nature of verification, which is neither a scientific nor a technical, but a social and cognitive process. Thus, the validity of mathematical proofs is based on the consensus of the members of the mathematical community that certain logical chains of reasoning lead to given conclusions. This means for the application to safety-related systems, and considering their importance for human life and health, but also for the environment and capital investments, that this consensus ought to be as broad as possible. Therefore, systems must be simple and appropriate program verification methods must be as generally understandable as possible—without sacrificing rigour.

The development of methods to check the safety of software is not very advanced, yet. So far, formal correctness proofs can only be carried out for relatively small program units—which fortunately cover already many safety-critical functions. Other rigorous methods that can often be used in this area are symbolic program execution and even complete tests. In Table 6, to each of the four safety integrity levels according to IEC 61508 [4] suitable programming paradigms and correctness verification methods are assigned in accordance with the respective requirements for the trustworthiness of the results to be achieved.

The following four sections present subsets of PEARL (see Fig. 5) assigned to the safety integrity levels SIL1 to SIL4, each of which limited to the bare essentials, as simple and easy to understand as possible, and nested within one another. Their syntactic form follows the tradition of PEARL, i.e. the formalisation is so little that even these language constructs can be read immediately and interpreted unambiguously by engineers without prior training. They are defined in such a way that the correctness of programs formulated in a sublanguage can be verified using the verification method corresponding to it in accordance with Table 6. The level SIL4 corresponds to the highest safety requirements. The higher the safety integrity level, the less flexibility is permitted in program formulation. Less safe language features are gradually prohibited at higher levels,

Table 6. Assignment of verification and programming methods to the safety integrity level of IEC 61508

Safety integrity level	Verification method	Programming method
SIL4	Social consensus	Cause-effect table
SIL3	Diverse back translation	Function block diagrams based on verified libraries
SIL2	Symbolic execution Formal correctness proofs	Procedure call Assignment Selection of alternatives Repetition-restricted loop
SIL1	All	Inherently safe, static, user-oriented

which is why it is not necessary to learn a new language for each safety integrity level, and compilers can check whether programs meet certain safety requirements.

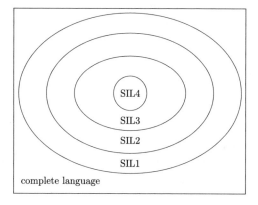

Fig. 5. Nesting of the PEARL subsets

The principle of defining subsets of a language for critical applications makes it possible to develop programs according to specific safety requirements, and to seamlessly link code sections for safety-critical and non-safety-critical system parts. The more safety-critical a system is, the more restrictive methods must be used. To enforce this, the desired safety integrity level **SIL1** to **SIL4** or, if there are no safety requirements, **NONE** is stated to compilers using the directive **SAFEGUARD**. This allows for an automated compile-time check to ensure that the specified language subset is adhered to. In a code segment, only constructs of the language subset of the specified and of higher safety integrity levels are permitted. Code segments with different safety requirements can be nested so that they can be transparently linked together.

4.2   SIL1: Constructive Preclusion of Error Sources

In safety-related automation programs, many variables denote physical quantities. However, the relations between their numerical values and the corresponding physical dimensions solely exist in the perception of the programmers. Mismatching units in different program components are a significant source of errors, which have already led to rocket crashes and the loss of satellites. For this reason, annotations are defined from which compilers do not generate executable code, but with which they can check the correct use of physical dimensions and, thus, consistency.

Bearing in mind that constructive preclusion of possible errors is the most effective measure to fulfill safety requirements, at the lowest safety integrity level SIL1 unconditional jumps, the use of pointers and references as well as unstructured synchronisation with semaphores are no longer permitted as the most error-prone language constructs.

Semaphores are implementation-oriented procedural constructs at a very low level, but completely inappropriate at the application level. In their place, functional synchronisation means are required which, when handling mutual exclusions, make it very clear which resources are to be protected and how. Furthermore, it must be possible to specify maximum waiting times before entering critical regions, maximum retention times in these regions, and suitable exception reactions. The monitor, barrier, mailbox and rendezvous synchronisation tools provided in other programming languages do not adequately meet these requirements. The **LOCK** instruction with temporal monitoring functions is, therefore, introduced to ensure secure and deadlock-free access to resources and to enforce their release after use:

Lock-statement ::=
    LOCK synchronisation-clause
    [waiting-time-clause] [execution-time-clause] [**NONPREEMPTIVELY**]
    PERFORM statement(s)
    UNLOCK ;

When a task executes such a lock instruction, the task is put on hold until all shared variables or data stations mentioned in the synchronisation clause can be seized in the specified manner, i.e. either for exclusive or shared access. By providing the optional waiting time clause, the waiting time before entering the lock instruction can be limited. An exception reaction can be specified to handle the case that the lock cannot be executed in the specified time frame. If locking is successful, the **PERFORM** instructions are executed. To ensure deadlock-free operation, a hierarchical order can be defined on the objects to be shared. Nesting of lock instructions is then permitted as long as the sequence of requests for further resources matches this order. If the lock instruction is terminated normally or left prematurely, the corresponding releases are made. Individual shared objects can, however, also be explicitly released prematurely. The optional execution time clause limits the time during which the executing task may remain in the critical region defined by the lock instruction. Exceeding the execution time leads to a system exception. Finally, the purpose of the optional attribute **NONPREEMPTIVELY** is to prevent the surrounding task from being pre-empted by the operating system for reasons of the applied processor allocation strategy during the execution of the lock instruction.

4.3   SIL2: Predictable Time Behaviour

Only very rarely, the full extent of higher real-time programming languages with asynchronous multitasking is required to formulate automation functionalities. An inherently safe sublanguage of PEARL is, therefore, defined in accordance with the requirements of SIL2, limiting itself to what is really necessary. In favour of reliability, it dispenses with all constructs that could lead to unpredictable capacity and runtime requirements. Thus, explicit runtime limits must be provided for procedure executions, exception handling and message exchange functions in distributed systems, as well as appropriate reactions in the event of timeouts. The same applies to repetition instructions. If the number of loop body executions exceeds the limit specified for this, loop processing is terminated and a specified statement sequence is executed, after which the control flow branches to the first statement behind the loop.

The priority concept used so far to determine execution sequences in situations of competition for resources is dependent on context, environment and implementation and is, therefore, not adequate. In contrast, users actually want to be able to set completion dates for system reactions at the specification level. In order to enable the use of timely strategies for preemptive and non-preemptive dynamic allocation of asynchronous computing processes such as earliest-deadline-first scheduling [2], upper time limits for the required execution times must be specified in appropriately extended task declarations, or activation or deferment instructions. For this purpose, either time durations are explicitly given, or the compiler is instructed to derive upper estimations of the maximum remaining task runtimes. If these limits for maximum permitted runtimes are exceeded, the operating system aborts the corresponding tasks and initiates exception handling. Completion dates for task executions are also specified in task declarations or activation, continuation or deferment instructions, either in form of absolute times or of deadlines. Is in the latter case a task's condition for (re)activation fulfilled, its completion date is determined by adding the specified deadline to the current time. If completion deadlines are violated, the operating system also terminates the tasks concerned and handles the corresponding exceptions.

The application-oriented option of setting deadlines for task processing entails the problem of missed deadlines and overload. In the interest of predictability and reliability of execution behavior, it is necessary to recognise and handle such exceptional situations as early as possible. Deadline-driven scheduling [2] checks every time a task enters the ready state, whether the set of ready-to-run tasks can be processed on time and, thus, recognises imminent overload situations. To handle the latter, only tasks marked accordingly in their declarations are left in the set of ready-to-run tasks, while all other ones are terminated and emergency processes are activated if necessary.

In order to be able to prove, with the help of tools, the correctness of programs formulated in this sublanguage using formal methods such as the Hoare calculus, their applicability was made possible by introducing assurances. To any procedure permitted at the safety integrity level SIL2 preconditions and postconditions as well as invariants in the form of Boolean expressions are assigned in its declaration. The preconditions are checked before execution, and the postconditions afterwards. The invariants must be valid both before and after the procedure is executed. If conditions are violated, an exception signal is generated.

4.4   SIL3: Function Block Diagrams

The graphical Function Block Diagrams defined in the international standard IEC 61131-3 [3] represent a programming paradigm that leads to both easily comprehensible and verifiable source and object code. With its long tradition in control engineering, graphical programming in the form of function block diagrams is well established in automation technology. Function block diagrams only know the four structural elements function and function block instances, data flow lines, identifiers and (external) connection points (cp Fig. 6). Accordingly, a partial language is introduced in PEARL for the textual implementation of the function block diagram paradigm in accordance with the requirements of SIL3, which mainly comprises just invocations of already verified procedures and parameter transfers.

This approach makes it possible to develop programs equivalent to function block diagrams very simply in the two steps of once-in-a-lifetime creation of a function (block) library and application-specific linking of function (block) instances. Accordingly, programs constructed as function block diagrams are also verified in two stages. Before a library is released, all the functions and function blocks it contains are verified using suitable methods as part of a type approval. Such a rather expensive safety licensing only needs to be carried out once for a specific application area after a suitable function (block) set has been identified. The licensing costs justified by safety requirements can, thus, be allocated to many implementations. Normally, rather few library elements are sufficient to formulate all programs in a specific area of automation technology. Then, for any given application program, only the correct implementation of the corresponding interconnection pattern of invoked functions and function block instances, i.e. a specific data flow, needs to be verified. For this purpose, the generally understandable method of diverse back translation [5], which can be used easily and economically by automation engineers and software testers, can be employed at the interconnection level of programs assembled of already verified function (blocks). Such programs have the quality and level of application-oriented specifications.

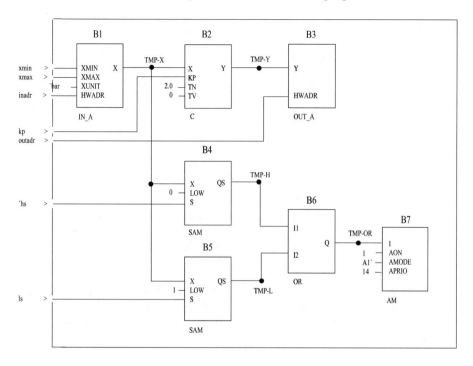

Fig. 6. Function block diagram of a program for pressure control and monitoring

4.5   SIL4: Cause-Effect Tables

A programming paradigm suitable for the highest safety integrity level SIL4 has long been well established in industry for the design of emergency shutdown and other protective systems. Software for protective systems is represented in form of cause-effect tables (cp. Fig. 7) due to the clarity and unambiguity that can be achieved and, thus, has the quality of specifications of certain functionalities. Hence, specifications and executable programs are identical. A syntax compatible with PEARL is defined for the textual formulation of such tables in a simple, clear and generally understandable, but nevertheless exact and unambiguous formal manner. Using linguistic expressions, this sublanguage defines which effects occurring causes should have. Since cause-effect tables can be easily checked and verified by the widest possible social consensus, they are the ideal form to formulate control systems meeting the highest safety requirements. The functionality thus provided at safety integrity level SIL4 is rather low, but completely sufficient to be able to formulate control systems of the highest safety criticality in a high-level language.

The rows of cause-effect tables are associated with events whose occurrence causes logical preconditions. In such a table, by placing marks in a specific column which is associated with an action, preconditions are selected that must all be fulfilled in order to trigger the action's execution. Surrounded by a special form of the module construct, in which the required process inputs and actuator

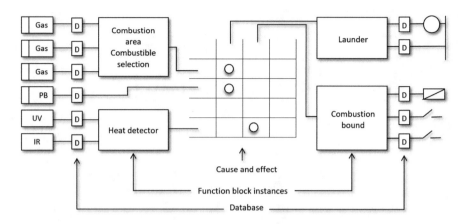

Fig. 7. A typical cause-effect table

outputs are specified, the new language construct **CETABLE** maps cause-effect tables textually into the form of logical conditions. If such a logical condition is fulfilled, the instruction associated with it is executed. The main syntax rules of this construct read:

Cause-effect-table ::=
    CETABLE CET-identifier [SIL-specification] ; [declarations]
    CET-line(s)
    END CET-identifier;
CET-line:=
    CAUSE expression of type **BIT(1)**
    EFFECT { **CALL**-statement | **SEND**-statement } ;

They are illustrated by the following example of a boiler, whose outlet valve is to be opened or closed depending on the pressure and temperature in the boiler, and the filling level of the liquid in it.

```
MODULE(Cause-Effect-Table) SAFEGUARD SIL4;
SYSTEM;
SPECIFY PressureSensor, FluidLevel, Thermocouple
        DATION IN SYSTEM BASIC;
SPECIFY Valve DATION OUT SYSTEM BASIC;
PROBLEM;
CETABLE Boiler;
    DECLARE Pressure, FillingLevel, Temperature FIXED;
    TAKE Pressure FROM PressureSensor;
    TAKE FillingLevel FROM FluidLevel;
    TAKE Temperature FROM Thermocouple;
    CAUSE Pressure LE 5 AND FillingLevel LE 85 AND Temperature LE 44
        EFFECT SEND '0'B1 TO Valve ;
    CAUSE Pressure LE 5 AND FillingLevel LE 85 AND Temperature GT 44
        EFFECT SEND '0'B1 TO Valve ;
```

```
    CAUSE Pressure LE 5 AND FillingLevel GT 85 AND Temperature LE 44
          EFFECT SEND '1'B1 TO Valve ;
    CAUSE Pressure LE 5 AND FillingLevel GT 85 AND Temperature GT 44
          EFFECT SEND '0'B1 TO Valve ;
    CAUSE Pressure GT 5 AND FillingLevel LE 85 AND Temperature LE 44
          EFFECT SEND '0'B1 TO Valve ;
    CAUSE Pressure GT 5 AND FillingLevel LE 85 AND Temperature GT 44
          EFFECT SEND '1'B1 TO Valve ;
    CAUSE Pressure GT 5 AND FillingLevel GT 85 AND Temperature LE 44
          EFFECT SEND '1'B1 TO Valve ;
    CAUSE Pressure GT 5 AND FillingLevel GT 85 AND Temperature GT 44
          EFFECT SEND '1'B1 TO Valve ;
END Boiler;
MODEND Cause-Effect-Table;
```

4.6   Synopsis of the Language Subsets

The nested subsets of PEARL for each of the safety integrity levels SIL1 to SIL4 according to IEC 61508 are summarised in Table 7. They were formed in such a way that the use of less safe language constructs is restricted more and more with increasing safety integrity levels, starting from applications without safety relevance ("NONE"). This provides a real-time programming language which comprises all language tools known to foster functional safety, and which is oriented towards human comprehension. Viewed as a social process to achieve consensus, program verification is facilitated by features such as composition and reuse of licensed components, programming at the specification level by creating cause-effect tables and, generally, by striving to achieve utmost simplicity in all aspects. The certification institutions can examine software written in the language with greater confidence and reasonable effort. Using the language to develop embedded systems promises to reduce the risk to human life, the environment and equipment, as well as maintenance costs, because the inherently safe language constructs mean that fewer mistakes are made in the first place.

5   Safe Sequential Function Charts

The language Sequential Function Chart according to IEC 61131-3 [3], which emanated from Grafcet, is not well suited for safety-related control tasks, as its syntax allows the programming of system deadlocks and violations of a number of safety rules to which sequential function charts are subject. In order to both remedy these shortcomings with as little effort as possible, and to provide in PEARL support for sequential function charts, which are the main tool to design software for programmable logic controllers, suitable extensions to the syntax of the language to be used up to safety integrity level SIL3 are defined. As a structuring element for main programs, procedures and processes, the concept

Table 7. Definition of safety-related language subsets

Language construct	NONE	SIL1	SIL2	SIL3	SIL4
unconditional jumps	+	−	−	−	−
(conditional) expressions and assignments	+	+	+	−	−
alternative and case selections	+	+	+	−	−
physical units	+	+	+	+	+
cause-effect tables	+	+	+	+	+
sequential function charts	+	+	+	+	−
synchronisation with semaphores	+	−	−	−	−
synchronisation with lock construct	+	+	+	−	−
internal signals	+	+	+	−	−
interrupt signals	+	+	+	−	−
asynchronous multi-tasking	+	+	+	−	−
with priorities	+	+	−	−	−
with deadlines and temporal monitoring	+	+	+	−	−
(function) procedure calls	+	+	+	+	−
repetitions	+	+	−	−	−
with limitation of runs	+	+	+	−	−
pointers and references	+	−	−	−	−
formatted and computer-internal I/O	+	+	+	−	−
process data input und output	+	+	+	+	+
distributed systems	+	+	+	+	+
dynamic reconfiguration	+	+	+	+	−
message transfer	+	+	+	−	−

+ permitted   − not permitted

fits easily into the existing PEARL syntax. This was achieved by the significant deviations from the syntax of sequential function charts according to the IEC 61131-3 standard described below, but without changing the functionality.

Any sequential function chart is encapsulated in its entirety by syntactic brackets. Particularly distinguished initial steps can be dispensed with, since the syntax enforces the existence of a first step at the beginning of each chart. The structure of sequential function charts or the corresponding directed graphs is reproduced by nesting the new language elements, and made clear by explicitly using a separate construct to select flow alternatives. This makes the naming of steps provided for in IEC 61131-3 redundant. There, also actions are given names to associate them with steps. This and the possibility of explicitly declaring actions seem to be redundant as well. Therefore, in the syntax presented here, the actions associated with a step are listed directly in the bodies of the steps in form of sequential code, procedure calls and tasking instructions and are, thus, linked to them. The qualification of actions also proves to be unnecessary, as PEARL already provides comprehensive capabilities for process and time control, which can be employed in the step bodies, and with which the same effects can be achieved.

Finally, only one form of branching to alternative sequences to be selected is provided, because the effect of the possibility of explicit prioritisation of the individual alternatives provided for in IEC 61131-3 can also be achieved by reordering in the write-up. In accordance with the requirements of the standard IEC 848, in IEC 61131-3 it can optionally be specified that the individual transition conditions initiating the various alternative branches should be mutually exclusive. It remains the programmer's task to ensure this property of the conditions. To prevent a possible source of error, this option is also omitted in the PEARL extension.

To prevent inconsistent behaviour, the new language constructs ensure that sequence controls always end with a step. For safety reasons, parallel branches, i.e. concurrency, and the possibility of programming sequence cycles are deliberately omitted, as the semantics of sequential function charts provide for the cyclical execution of the individual steps anyway. However, sequential function charts are often processed repeatedly in their entirety. This can easily be achieved in PEARL by embedding a chart in a loop construction. Statically planned cyclical execution or execution organised by sequential function charts is the only form to schedule computing processes permitted above safety integrity level SIL1.

The additional language constructs for PEARL to formulate sequential function charts are now defined according to the above considerations by the production rules given in extended Backus-Naur form. The character | is used to list selectable options, that can be combined into groups with (), and [] is used to identify options, while { } denotes at least one repetition. The newly introduced keywords are self-explanatory. The syntax rules enforce alternating sequences of steps and transitions and the properties mentioned above.

SFC::= **SEQUENCE** SFC-body **ENDSEQ**;
SFC-body::= Step [(Transition | Alternatives) SFC-body]
Step::= **STEP** [Statement(sequence)] **ENDSTEP**;
Transition::= **TRANSITION** expression of type **BIT(1)**;
Alternatives::=
    SELECT {**BRANCH** Transition SFC-body Transition} **ENDSCT**;

As an example, Fig. 8 shows the sequential function chart of a boiler automation system, which consists of a boiler to be filled on request by a touch signal, its contents heated and then emptied. This flow chart is formulated in PEARL as follows:

```
SEQUENCE
    STEP ENDSTEP;
    TRANSITION Digital_in(adr) AND Temperature LE tmin;
    STEP Call Fill ENDSTEP;
    TRANSITION FillingLevel GE lmax;
    STEP Call HeatUp ENDSTEP;
```

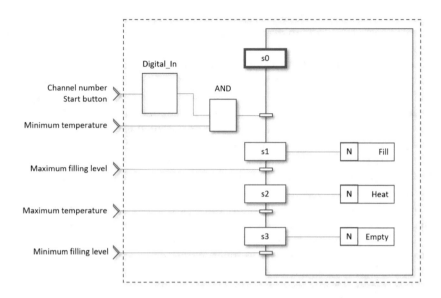

Fig. 8. Sequential function chart of a boiler automation

```
        TRANSITION Temperature GE tmax;
        STEP Call Empty ENDSTEP;
        TRANSITION FillingLevel LE lmin;
        STEP ENDSTEP;
ENDSEQ;
```

6   PEARL as Specification Language

The prevailing practice in real-time programming is characterised by the use of unsuitable tools. In most cases, even languages without real-time capabilities are employed for real-time applications, forcing programmers to compensate for the shortcomings of these languages in complicated ways, which are difficult for third parties to understand and not portable, by means of operating system calls, assembler inserts and the like. This situation is completely incomprehensible, as genuine real-time programming languages such as PEARL with comprehensive, application-oriented expressiveness are already existing for a long time.

There are two reasons why it makes sense to use an—extended—subset of PEARL prior to the actual programming: First, to introduce developers of real-time applications to a genuine real-time programming language, and secondly, to supplement the process of software construction and validation with explicit formulation of requirements and clear representation of the system architecture. Although syntactically correct language constructs should be used for this, no executable PEARL programs are to be generated. Instead, program specifications are made up of texts, in which the software structure is described using specific language constructs, but details and algorithms are specified by comments in natural language. On the one hand, due to the utilisation of application-oriented notions, the suitability of PEARL for this purpose is based on its inherent documentation value and the precise notation of PEARL code, but also on its easy readability for clients, developers and users, and, on the other hand, on the existence of language constructs that go far beyond the scope of other programming languages.

Now those language constructs already available in the normal extent of PEARL, which meet the requirements as means of expression for specification purposes, are explained in more detail. All the language elements listed are extremely well adapted to the needs of automation technology, and their syntax reflects the terms essential for distributed real-time applications in an easily readable form. PEARL is therefore suitable both as a programming language and as a specification language, especially for engineers.

Modularisation. To structure software, PEARL supports a consistent modularisation concept in a particularly clear form. This not only comprises the actual software modules, but also the additional structuring options provided by system and problem parts, computing processes and procedures, which serve to differentiate between software and other system components as well as to include relevant influencing quantities of the system environment. At a very early stage in the software development process, dedicated software modules can, in this way, be biuniquely associated with each subprocess in technical processes to be automated. This simultaneously prepares the way for a division of labour in software development and, if necessary, geographically distributed implementations. Written in PEARL, initially modules are relatively meaningless hulls whose intended functionality is solely defined by intuitively understandable technical terms and short natural language comments. As the development process progresses, these terms are then gradually refined and supplemented. This process ultimately leads to valid, executable PEARL program text in the course of the actual implementation.

Device Connections and Interconnections. Planning, installation and management of process input and output equipment play a major rôle in automation technology. The associated documentation is generally very extensive and requires considerable effort to maintain. Clear descriptions of all process interfaces are a fundamental prerequisite for program creation and must, therefore, be provided during the specification phase. PEARL also differs from other real-time programming languages in that the description of the interfaces to the outside world is an indispensable part of PEARL programs. This also holds for the languages according to IEC 61131-3 [3]. For reasons of portability, these descriptions are concentrated and encapsulated in system parts. It is a considerable simplification of work, being able to make the necessary descriptions of device connections and interconnections readily in form of PEARL system parts. Owing to PEARL's easy readability, the formalism to be adhered to does not represent a restriction with regard to the use of such interface documentation for different purposes and by different user groups. Additional information not required by PEARL can be included in system parts as comments.

External Interrupt Sources. The use of the interrupt concept in automation programs poses a number of risks in terms of clarity, comprehensibility, predictability and reliability of system behaviour. For this reason, clear documentation of all interrupt sources effective in a program system is of fundamental importance. This requirement for good software specification is already taken into account by the normal PEARL language scope, too. Interrupts must be described in the same way as other process inputs in system parts.

User-Defined Data Stations. With the concept of the data station (DATION), PEARL offers a language construct allowing developers to specify the characteristics of complex peripheral devices or input and output interfaces as sources and sinks of bit streams, which are relevant from the user's point of view, and are required for programming, completely independent from implementation. Such specifications provide clear information about which functionalities the developers of application programs can expect on the one hand, and which services the corresponding driver programs must provide on the other.

Concurrency. The concept of computing processes is an important element to structure program systems. In principle, it is implementation-independent and is generally used to formulate event-dependent activities. In other words, process objects are the basic elements of reactive systems. They specify the functionality of the reactions, but not how these are ultimately achieved, e.g. by means of strictly sequential, concurrent or physically parallel execution. During the specification phase, the developers may therefore model all dependencies between individual tasks contained in a problem as such, but are not forced to over-specify such tasks through unnecessary sequentialisations and, thus, to unnecessarily restrict the room for implementation decisions.

Scheduling Conditions. The events triggering the execution of tasking operations, in particular process activations, are described in PEARL by scheduling conditions. These are the occurrences of interrupt signals as well as individual and periodically repeating points in time. Either in natural language or in another form it is specified which reaction, in form of executing one or more processes, is expected when an event occurs. Here too, formulation directly in PEARL is a good option, especially as the syntax of scheduling conditions is very similar to plain text.

Distributed Systems. The above explanations show that PEARL for single-processor systems already has a whole range of language elements, which can be useful for system specification. For distributed systems, it is actually more a specification language than a programming language. Indispensable for specifying the behaviour of distributed systems, it allows to

- describe the hardware configuration,
- describe the software configuration,
- specify communication and its properties (peripheral and process connections, physical and logical interconnections, transmission protocols) and
- specify the conditions and the implementation type of dynamic reconfigurations in the event of failure.

In contrast, there are only a few executable language constructs in the classic sense, namely for exchanging messages. Even if PEARL is not employed for programming, it ought to be used to specify structure and behaviuor of distributed systems, because practically no other language offers these capabilities in a way as easy to read and understand.

As an example of employing selected PEARL language elements for specification purposes, we consider the heating regulation for two apartments described in detail and fully PEARL-coded in [6]. The corresponding automation program is combined in one module. Of the system part, which is used exclusively for specification purposes, only the description of two interrupt signals is mentioned here:

```
Failure : Hardware address where the failure signal is connected
Terminal: Hardware address of user terminals
```

The sequence control of the heating regulation is then described in "Specification PEARL" as follows:

```
/* Initialisation of the particular regulation parameters */
WHEN Failure  ACTIVATE Disturbance_Logging_and_Removal;
WHEN Terminal ACTIVATE Interactive_Temperature_SetValue_Input;
ALL Sampling_Interval_Apt_1  ACTIVATE Temp_Regulation_Apt_1;
ALL Sampling_Interval_Apt_2  ACTIVATE Temp_Regulation_Apt_2;
ALL Sampling_Interval_Boiler ACTIVATE Temp_Regulation_Boiler_Water;
```

Owing to the unambiguousness of precisely defined scientific and technical terms, a verbal description of algorithms is very often sufficient to clearly specify a certain functionality. In our example, we could write:

```
/* The temperatures in the apartments are    */
/* regulated with a discretised PID algorithm. */
/* The water temperature in the boiler is     */
/* regulated with the two-point algorithm.    */
```

References

1. DIN 66 253: Programmiersprache PEARL – SafePEARL. Beuth Verlag, Berlin-Cologne (2018)
2. Henn, R.: Feasible processor allocation in a hard-real-time environment. Real-Time Syst. **1**(1), 77–93 (1989)
3. IEC 61131-3: Programmable Controllers, Part 3: Programming Languages. Edition 3.0. International Electrotechnical Commission, Geneva (2013)
4. IEC 61508: Functional safety of electrical/electronic/programmable electronic safety-related systems. Part 1: General requirements. Edition 2.0. International Electrotechnical Commission, Geneva (2010)
5. Krebs, H., Haspel, U.: Ein Verfahren zur Software-Verifikation. Regelungstechnische Praxis RTP **26**, 73–78 (1984)
6. Lauber, R.: Prozeßautomatisierung, Band 1, 2. Auflage. Springer, Berlin-Heidelberg-New York-London-Paris-Tokio (1989)

Meta-Learning for Time Series Analysis and/or Forecasting: Concept Review and Comprehensive Critical Comparative Survey

Witesyavwirwa Vianney Kambale, Denis D'Ambrosi, Paraskevi Fasouli, and Kyandoghere Kyamakya$^{(\boxtimes)}$

Institute for Smart Systems Technologies, Universität Klagenfurt, Universitätsstraße 65/67, 9020 Klagenfurt, Austria
kyandoghere.kyamakya@aau.at

Abstract. Meta-Learning has emerged as a solution to address the limitations of data unavailability and the lack of extensive computing resources. The aim of this work is to consolidate a discussion on the application of Meta-Learning in Time Series forecasting, specifically by comprehensively contrasting Zero-Shot Learning (ZSL), One-Shot Learning (OSL), and Few-Shot Learning (FSL). In our implementation setup, the comparative analysis of results identifies Few-Shot Learning as the best performer and One-shot learning as the worst. The performance improvement registered for FSL is credited to the additional meta-knowledge learned through the MAML-based method. Implemented as a Siamese network, OSL had to learn the strong periodic components within the time series. However, the selected datasets did not display such strong periodicity; instead, they were dominated by trend and noise components. Future work intends to analyze the asymptotic performance of increasingly complex predictors in tandem with increasingly longer training regimes.

1 Introduction

In this section, we provide a background and motivation for Meta-Learning in Time Series Analysis and Forecasting. We then formulate the research questions for the study at hand.

1.1 Background and Motivation

Over the past few decades, deep learning (DL) has garnered significant attention within the research community due to the exceptional performance of DL algorithms across a variety of tasks. Specifically in the domains of Time Series Analysis (TSA) and Time Series Forecasting (TSF), DL has been acclaimed for its ability to automatically learn temporal dependencies and handle complex, non-linear relationships within data. Furthermore, DL algorithms have demonstrated remarkable efficacy in diverse Time Series Forecasting tasks [2].

H. Unger and M. Schaible (Eds.): AUTSYS 2023, LNNS 1009, pp. 80–109, 2024.
https://doi.org/10.1007/978-3-031-61418-7_4

However, the successes of DL have predominantly been observed in areas with substantial data availability and where significant computing resources are accessible [3]. Consequently, various applications where data is inherently scarce or costly to acquire, as well as scenarios lacking extensive computing resources, have been ruled out from leveraging the advancements of DL. For instance, even though the burgeoning of 5G technology results in the generation of a substantial amount of data by end devices, the size of individual data sets can be notably small [1]. Traditional deep learning (DL) methods, which rely heavily on large data sets, underperform in scenarios characterized by limited samples. In such contexts, Meta-Learning, particularly in the form of Few-Shot Learning, emerges as a promising approach to address these data scarcity challenges.

Meta-Learning, a paradigm that leverages prior knowledge to enhance learning efficiency, has emerged as a solution in addressing the limitations due to data unavailability and lack of extensive computing resources [4]. Referring to the concept as "learning to learn", Meta-Learning aims to improve a learning algorithm over multiple learning episodes.

Meta-Learning applications have gained prominence in the field of computer vision [5]. However, in the domain of Time Series Forecasting, research remains relatively sparse [6,7]. This work aims to consolidate a discussion on the application of Meta-Learning to Time Series Forecasting, specifically comprehensively contrasting Zero-Shot Learning, One-shot learning, and Few-Shot Learning.

1.2 Problem Statement and Research Questions

From the discussion in Sect. 1.1, it is evident that comprehensive works providing a condensed analysis of aspects of M-L with a specific focus to Time Series Forecasting are needed. We therefore formulate the following research questions.

RQ1: How do Meta-Learning approaches for Time Series Analysis and Forecasting fit within the existing Meta-Learning frameworks and taxonomies? To answer this question, we begin by reviewing the fundamental concepts of Meta-Learning. Then, we discuss the classification and regression problems as applied in Meta-Learning for time series data. Finally, we provide a brief overview of the Meta-Learning approaches and taxonomies found in the literature, discussing how Meta-Learning for Time Series Analysis and Forecasting integrates into this landscape.

RQ2: What is the state of the art (SOTA) in Meta-Learning for Time Series Analysis and/or Forecasting with respect to Zero-Shot Learning (ZSL), One-Shot Learning (OSL), and Few-Shot Learning (FSL)? To address this question, we first establish the relevance of Meta-Learning in Time Series Analysis and/or Forecasting. Then, we formulate Meta-Learning problems for ZSL, OSL, and FSL, covering both classification and regression tasks. Next, we critically discuss the concepts of ZSL, OSL, and FSL within the context of Meta-Learning. This discussion sets the groundwork for a comprehensive review of the SOTA in Meta-Learning with respect to ZSL, OSL, and FSL for Time Series Analysis and/or Forecasting. Finally, we present a critical comparative analysis of Meta-Learning with respect to ZSL, OSL, and FSL in the context of Time Series Analysis and/or Forecasting.

RQ3: How can we empirically assess the efficacy of Zero-Shot, One-Shot, and Few-Shot Learning techniques in Time Series Forecasting? This question is addressed by critically evaluating the performance and applicability of Zero-Shot Learning, One-Shot Learning, and Few-Shot Learning methods in the context of Time Series Forecasting.

The rest of this work is organized as follows: Sect. 2 discusses the fundamentals of Meta-learning. In Sect. 3, we present a comprehensive review of Meta-Learning in Time Series Analysis and Forecasting with regard to Zero-Shot Learning, One-Shot Learning and Few-Shot Learning. Section 4 presents illustrative implementations of ZSL, OSL and FSL. Section 5 discusses the challenges and limitations of current Meta-learning techniques, as well as directions for future research in the context of Time Series Analysis and Forecasting. Finally, concluding remarks are provided in Sect. 6.

2   Fundamentals of Meta-Learning

In the broadest sense, the Supervised Machine Learning (ML) problem can be described as approximating an arbitrary unknown function, given a dataset of observations that contain (ideally) numerous mappings of said function, by updating an initially randomly behaved model through one or multiple iterations over the data. Even though this definition encompasses a wide umbrella of algorithms, the update process (also known as *learning*) generally consists in a carefully constructed mechanism that improves the performances of a predictor $\mathcal{P}_\theta$ by modifying its internal parameters θ in order to minimize a given loss function $\mathcal{L}$, computed between the ground truths belonging to the dataset $\mathcal{D}_y^{(tr)}$ and the forecasts produced by $\mathcal{P}_\theta$ on the input datapoints $\mathcal{D}_x^{(tr)}$. The task that the predictor is training on can be expressed as a pair $\mathcal{T} = (\mathcal{D}, \mathcal{L})$, where $\mathcal{D} = (\mathcal{D}^{(tr)}, \mathcal{D}^{(test)})$ and $\mathcal{D}^i = (\mathcal{D}_x^i, \mathcal{D}_y^i)$, $i \in \{(tr), (test)\}$. The said process can be formulated as in Eq. 1.

$$\theta' = \underset{\theta}{\operatorname{argmin}} \ \mathcal{L}\left(\mathcal{P}_\theta(\mathcal{D}_x^{(tr)}), \mathcal{D}_y^{(tr)}\right) \tag{1}$$

In this setting, we suppose that there are several hyperparameters that influence the training procedure, like the structure of the predictor, its initialization, or the choice of optimizer, but they are determined at the beginning of the experiment and do not evolve as new observations are provided.

Even though Machine Learning has provided great value in the modern data-driven society, it still poses an efficiency problem: models are typically trained from scratch, without leveraging past training experience, thus requiring huge datasets and extensive time and energy resources. To overcome this issue, the idea of Meta-Learning (M-L) emerged. Derived directly from cognitive sciences, Meta-Learning is a subfield of Artificial Intelligence that studies how the training process of ML predictors can be sped up and cost-amortized by exploiting meta-information derived from previous training tasks. It is important to highlight that M-L is not a technology in itself, but rather a set of techniques that can be applied in correlation with a (more or less) specific ML problem to enhance the final output, both performance and efficiency-wise. Whereas in Machine Learning we have a model that learns the features of a function from examples extracted from the distribution of the training dataset $\mathcal{D}^{(tr)}$, in Meta-Learning a Meta-Learner tries to generalize the characteristics of the learning procedure itself through multiple episodes of training across multiple datasets taken from a fixed task distribution. Meta-Learning aims at optimizing the hyperparameters of the inner *base* learner to streamline its training process, thus allowing it to adapt more quickly to new scenarios. More abstractly, M-L can be formalized as a bilevel optimization problem, where the outer Meta-Learning objective is to learn the hyperparameters searched through multiple runs of the training procedure described in Eq. 1 on different datasets, such that the predictor will be facilitated during the training on a new, unseen task. For further reference to the formalization of Meta-Learning, we refer to [3].

2.1   Fundamentals Concepts of Meta-Learning

In this section, we will delve deeper into the idea introduced above, presenting the main lexicon related to Meta-Learning.

The key concept is the Meta-Learner: it could be either a deterministic algorithm or an entire ML model whose objective is to collect and exploit meta-data from "standard" ML training procedures to facilitate rapid adaptation. Meta-Learners can either be trained on families of tasks that share similarities in structure to learn dataset-agnostic features, or on the same task multiple times to improve the ML model performances: hyperparameter tuning can be considered, in fact, a form of Meta-Learning.

Whereas a classical ML model gets trained and tested with two different partitions of the same dataset, a Meta-Learner treats a whole task as a "Meta-Learning" example and can potentially be tested onto completely unseen datasets. The Meta-Learner essentially "learns to learn" by inferring patterns that transcend individual tasks, thus building better generalization capabilities.

Similarly to the distinction between a classical ML learner and a Meta-Learner, we can discriminate between tasks, made up of a database and an objective (usually in the form of a loss function), and meta-tasks, which consist of collections of tasks with a similar structure. Training a Meta-Learner on a meta-task distribution forces it to learn the common features of the datasets contained in it, thus effectively building an inductive bias that allows it to perform well on tasks not previously encountered.

Continuing the analogy between ML and M-L, tasks are extracted from a fixed distribution during Meta-Learning analogously to the way examples are sampled from a dataset during ML training. Each randomly sampled dataset exposes the Meta-Learner to a new scenario during an *episode* of training, enhancing its adaptability to unforeseen challenges. At the end of the episode, the model can be validated on a separate set of data either coming from already known or unseen tasks. Meta-testing, on the other hand, aims at measuring the adaptation capabilities of the Meta-Learner, and thus requires data collected from tasks that have not been previously sampled.

Meta-Learning adaptations can be forced in multiple ways. One very common strategy is called *fine-tuning*, which consists of training a ML model on a vast, diverse dataset to build a generalized internal representation for the data; in a later step, the same model gets optimized for a single task by training it on a much smaller and less heterogeneous dataset. This allows the model to adapt its pre-existent knowledge to the unseen datapoints, tailoring its predictions to the nuances of the new task, but without requiring as many resources as a training procedure done from scratch. This could be useful both in a scenario where we want to produce a series of specific models that work on a similar domain, akin to crafting a tailored recommender system, or if we needed to train a predictor in a situation of data shortage, a situation often encountered in anomaly detection.

Another common way of applying M-L techniques is to build higher-level abstractions of data coming from a particular field [8]: by extracting meta-features from a meta-task, we can encapsulate commonalities across a whole domain and, thus, have better generalization capabilities. One example of this strategy would be to construct a ML module capable of mapping inputs from different datasets to the same latent embedding space and then training a regressor/classifier model using this space as an input domain.

2.2 Meta-Learning for Time Series Data: Classification Problems vs. Regression Problems

Supervised Machine Learning has been repeatedly confirmed as a valid solution to a broad array of problems: in the past years it has been successfully applied to various domains such as stock forecasting, generative answering, image detection and more. Even though such heterogeneity may seem overwhelming at first, we can boil down all these tasks to two prototypical problems: classification and regression.

Classification tasks consist of partitioning a dataset of examples into a finite set of labels. For example, image recognition, sentiment analysis and spam filtering are all examples of classification problems. Even though these kinds of tasks typically rely on a heavily restricted and context-specific codomain, the generalization capabilities derived from the implementation of Meta-Learning techniques can provide great improvements to the performances of base learners trained on single-tasks datasets [9]. Especially in scenarios characterized by scarce or inherently balanced data, such as the field of anomaly detection, pre-training a model on a vast and varied dataset can be beneficial [10]. This initial training phase allows the base predictor to develop an efficient internal representation of the data that can be exploited later on during a fine-tuning procedure on a limited target dataset. Results, in particular coming from the image recognition field [11,12], show that base ML models require a lot of effort to learn to extract simple features that are generally shared across various tasks [13];

as a consequence, pretraining a model on a vast, heterogeneous dataset leads to intermediate parameter values that can be quickly optimized to discriminate between the details of a new dataset [14–16]. The subfield of ML that proposes such a set of techniques is often referred as to as *transfer learning* [17]. Another way to exploit Meta-Learning heuristics in the classification realm is to implement some form of prototypical network [18]. These models efficiently learn to construct latent prototypes for the classes encountered during meta-training and then classify new examples based on a similarity measure of their latent representation to each prototype within a latent space. Such techniques can be particularly helpful in dynamic domains, where new classes may emerge often, for example in the disease detection context, and retraining models from scratch each time would be cost-inefficient or straight-up unfeasible [20].

On the other hand, regression tasks feature a continuous, possibly infinite numeric codomain that examples must be mapped into. Examples of regression objectives could be estimating the position of a bounding box in an object detection algorithm, stock market forecasting or temperature prediction. Again, also in this field, Meta-Learning can improve a predictor's capabilities: for example, pre-training a Recurrent Neural Network on a big dataset of diverse time series can lead to better performances when forecasting a particular time series than only training it on a subset of that time series alone [19]. For a more comprehensive study on the application of Meta-Learning to the regression domain, we refer to [21].

2.3 Discussion of the Different Meta-Learning Taxonomies and Approaches

Since the general formulation of the Meta-Learning problem is quite simple and broad, "learning to learn" is a friendly and recurrent punchline used to describe the topic in almost any article about this subject. Therefore we need a robust and well-defined taxonomy to investigate this field in a structured way. Before the survey by Hospedales et al. [3], M-L techniques were generally classified into three uncomparable classes:

1. Methods that optimize hyperparameters through higher-order gradient-descent [22] or its first order approximations [23] (*optimization-based methods*).
2. Methods that, provided a training dataset, generate or initialize a model in an optimal initial configuration for quick learning on that particular dataset [24] (*black box/model-based methods*).
3. Methods that rely on comparing validation points with training points to produce a similarity metric (*metric-based methods* [25]).

The above survey updated this coarse taxonomy, introducing a more complete and understandable classification system. According to the author's proposal [3], Meta-Learning techniques should be discriminated along three, orthogonal axes: the *Meta-Representation*, the *Meta-Optimizer* and the *Meta-Objective*.

The Meta-Representation consists of the actual information that gets Meta-Learned. Possible examples of Meta-Representation could be, for example, loss functions [26], tuned parameters [27], embedding modules [28], initialization parameters [29], optimizers [30], model architectures [31] or data augmentation techniques [32].

The Meta-Optimizer is the concrete Meta-Learning technique adopted for meta-training. In practice, we can classify methods as either *gradient-based*, *reinforcement-learning-based* or *evolutionary-search-based*.

The Meta-Objective specifies which particular performance boost we are trying to obtain through the application of Meta-Learning. Some techniques may either aim at improving fast adaptability to new tasks with few examples [33], or asymptotic performance over a prolonged period of training with many datapoints [34], single [35] or multi-task [36] performance and feature online [37] or offline [38] meta-optimization. Implementing M-L techniques can lead to both efficiency and precision improvements, but since each heuristic inherently requires various trade-offs, it is important to thoroughly consider the various frameworks keeping in mind the nuances of the specific use case under analysis.

This three-axes taxonomy allows for fine comparisons between different Meta-Learning models and provides a good starting point to navigate through the maze of the literature of such a diverse and complex field. In the remaining part of this section, we will show how some of the seminal papers about Meta-Learning can be easily classified along the three axes.

Model-Agnostic-Meta-Learning (MAML) [38] is a highly referenced framework proposed in 2017 that aims at pretraining a model across a task distribution in order to provide quick convergence during future fine-tuning procedures. Given an untrained predictor $\mathcal{P}_\theta$, MAML (and analogous algorithms such as FOMAML [38], Sign-MAML [39] and Reptile [40]) tries to find the best initialization for θ (Meta-Representation), in order to provide great few-shot, multi task, offline adaptation (Meta-Objective), using either a gradient-descent-based or reinforcement-learning-based method (Meta-Optimizer).

In the article [41], the authors provide an automated method for discovering new loss functions for ML training customized for a predetermined task. The results show that ML models trained with cost functions (Meta-Representation) learned through evolutionary search (Meta-Optimizer) often lead to a quicker single task, offline convergence (Meta-Objective).

Finally, the idea behind [42] is to learn an augmentation method (Meta-Representation) for a target dataset, to asymptotically improve the classification performances of a predetermined model on it (Meta-Objective) through a reinforcement-learning-based method (Meta-Optimizer).

2.4   Critical Discussion of the Meta-Learning Taxonomies and Approaches with Focusing on Time Series Data

Although the taxonomy presented in Sect. 2.3 is domain-agnostic and, as a consequence, valid for any Meta-Learning implementation, it is imperative to introduce a domain-specific classification of the methods that allows for a finer comparison between the articles that deal with the specific challenges of time series.

We can further distinguish Meta-Learning approaches for Time Series ML based on:

- Number of input features
- Sampling flexibility
- Length of the history
- Explainability
- Length and variability of the prediction horizon
- Number of predictors involved

Most of Meta-Learning paradigms within the realm of Time Series ML are not designed specifically with multivariate distributions in mind, so they may require extensive modifications before being applicable to use cases in which we have to consider multiple features for each timestamp. On the other hand, other architectures belonging to both the classification [43] and the regression [44,45] domains, consider specifically the case of multivariate inputs and thus may be solutions quicker to deploy.

Meta-Training a predictor on a diverse set of time series, possibly coming from different domains, may lead to great generalization capabilities, but chances are that not all of the datasets available have been sampled with the same frequency. In such scenario, a model could either require limiting both the source and target datasets to a uniform sampling rate or adapt to varying frequencies, resulting in a potentially more versatile model. Furthermore, training a predictor on irregular patterns makes it more robust in case of missing datapoints, which can be caused for example by outlier removal or problems arisen during the sampling procedure.

Temporal data features two major characteristics that differentiate it from other types of structured information: one is the fact that datapoints are inherently ordered and sampled at different intervals that influence the trends of the values (see the previous point), and the other one is that time series may feature very different history lengths even when belonging to the same domain. Some architectures need a fixed length input, requiring cropping and padding operations that consequently cause some degree of information loss. Other models [45,46], instead, are more flexible and can accommodate time series with histories of varying sizes.

Explainability is still one of the main areas of concern for any complex ML model, above all in the case of modern Neural Networks, and Meta-Learning often introduces an additional layer of obfuscation. Nonetheless, there have been some recent interesting efforts towards developing partially interpretable Meta-Learning algorithms that provide current state-of-the-art performances [47,48]. By including explainability within the dimensions of our proposed taxonomy, we hope to highlight a significant unbalance in the existing research due to the predominance of uninterpretable models. By shedding light on this issue, we hope to ignite further research endeavors aimed at developing more transparent and explainable M-L frameworks.

Generally speaking, Time Series ML models are fed a history as input and produce a single value as output; this value could be a feature of the history (for example for detecting whether an ECG is showing signs of cardiac disease or not), or some information relative the next timestamp (for example forecasting the cost of a stock the next day given its evolution). Some Meta-Learning models, instead, can compute a prediction that spans multiple timestamps into the future or forecast a single value, but at a timestamp not determined a priori [45].

Finally, Meta-Learning architectures in this domain can be differentiated between single-predictor algorithms and ensemble methods. Contrary to the latter, ensembling consists of combining the predictions of multiple models into a more generalized, and less prone to overfit, forecast. Examples of such methodology applied to Time Series ML are [47, 49, 50]; interestingly enough, there already exists an up-to-date taxonomy of Meta-Learning ensembling methods for Time Series Forecasting [51].

3   Comprehensive Review of Meta-Learning in TSA/TSF with Regard to ZSL, OSL and FSL (RQ2)

To provide a comprehensive review of Meta-Learning in Time Series Analysis and Forecasting, this section starts by discussing the relavancy of M-L with time series data. The section then delves into providing the M-L problem formulations for ZSL, OSL and FSL and the succent pipelines for their implementation are discussed.

3.1   Relevance of Meta-Learning in TSA and TSF

As previously mentioned, Meta-Learning encompasses a huge umbrella of frameworks and techniques that aim at addressing different aspects of standard Machine Learning's shortcomings. Such diversity is also reflected when working with temporal data, as different strategies can overcome various challenges that may manifest themselves when dealing with this context. Note that from now on, we will refer to *Time Series ML* as both classification and regression tasks executed in this domain: although the techniques applied will be different, most of the challenges are agnostic to the actual objective of the model and depend on the very nature of time series themselves.

First of all, time series often feature dynamic patterns that may change suddenly. Effective ML models must be able to swiftly adapt to the evolution of trends while keeping information about seasonality to produce accurate predictions. Meta-Learning generally allows for better generalization capabilities that can easily accommodate unexpected changes in time series leveraging the knowledge derived from previous training episodes on other dynamic time series. Quick adaptability is a crucial *desideratum* for ensuring robust prediction capabilities even in dynamic scenarios and, as matter of fact, it is one of the most common Meta-Objectives in this field.

Not only time series values may feature unexpected patterns, also the sampling of the data may vary differently from case to case due to context restrictions. Some datasets may be constructed upon hourly, daily, weekly, monthly, yearly or unregular sampling, showcasing different widely different representations for possibly very similar trends. Datapoints may be missing because of circumstantial issues, temporal series may have vastly different lengths and scales and, for forecasting tasks, prediction horizons may vary. In summary, even though the input to any Time Series ML model may look fairly standardized, all the context information behind any single task will inherently condition the outcome of the predictions in a different way. Meta-Learning frameworks enable models to improve their generalization capabilities across the diversity of a series of support tasks, thus possibly leading to better performances when handling the particular nuances of the target datasets.

Furthermore, time series datasets are not exempt from all the data shortages problems that often plague all the other ML domains: producing a correctly labeled and, in the case of classification also balanced, dataset may not always be cost-effective or even possible. Recall the ongoing problem with anomaly detection: an occurrence, to be considered an anomaly, should occur very rarely by definition. Meta-Training a predictor on a distribution of similar tasks can possibly overcome this issue by providing zero-shot capabilities or allowing for less data-intensive fine-tuning.

In conclusion, although not previously investigated as in the domain of computer vision, Meta-Learning can be very relevant in the temporal scenario. Time series provide a unique set of challenges that may be successfully undertaken with both Meta-Learning ad-hoc techniques and model agnostic frameworks. In the following sections, we will delve deeper into the actual applications and use cases of Meta-Learning in this field.

3.2   Critical Discussion of ZSL, OSL and FSL in Meta-Learning

Even though the Meta-Learning realm is very vast and articulated, a good portion of the current research is oriented towards a single Meta-Objective: preparing models for fast adaptation. Solving this problem could provide solutions to many use cases: working around a shortage of data (for example when forecasting time series with limited histories), quick customization of ML predictors (for example for recommender systems that exploit information about a user's actions), or even reducing the carbon footprint of training ML models from scratch.

Often, such Meta-Objective is called *Few-Shot Learning* and is parametrized on the number of examples of the target dataset available for training:

1. *Zero-Shot Learning* means that we do not have any sample coming from the target distribution.
2. *One-Shot Learning* means that we have 1 sample for class in the classification setting (the case of regression is more delicate and we will explain why in the following section).
3. *K-Shot Learning* means that we have at least K samples for class in the classification setting and K samples in total in the regression setting.

Now we will explore the solutions for these scenarios considering the items one by one, but note that the set of methods capable of solving each particular

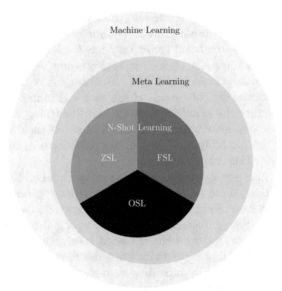

Fig. 1. Visualization of the inclusion relations between ML, M-L, ZSL, OSL and FSL

problem in the list is also clearly a superset of the techniques for the following one (see Fig. 1).

Zero-Shot Learning requires pretraining a model on data sampled from distributions similar to the target task, while generally ensuring that some mechanism is in place to ensure domain generalization. In a classification setting, it has been proven successful to consider auxiliary sources of semantic information (like textual descriptions in an image classification problem) to bridge the gap between seen and unseen classes [52]. For example, mapping visual features to semantic embeddings [53], semantic embeddings to visual features [54], or both semantic embeddings and visual features to a latent space [55], has previously allowed to perform classification on the codomain space via nearest neighbor search in a zero-shot manner thanks to the additional information. An alternative version of this idea that uses time series embeddings and statistic metrics instead of visual semantic features has been successfully applied to Zero-Shot Time Series classification [56]. Apart from devoting entire architectures to Zero-Shot Learning, there are also several domain-agnostic techniques which can be adopted by any classifier to improve the chances of recognizing classes unseen during meta-training, for example by decreasing the overall confidence of the model in any single class (*temperature scaling* [57]), or by selectively reduce its confidence for seen classes (*calibrated stacking* [58]). Even if there have been a lot of results with meta-embedding modules, they are not the only possible Meta-Representation that we can aim for. It is also possible to use generative models to create training examples for the unseen classes from a semantic description of their features [59]: such technique breaches into the realm of synthetic-data training, which is another key idea in Zero-Shot Learning. Models pre-trained on completely synthetic data and not fine-tuned in a subsequent stage are referred to as *Prior-Data Fitted Networks* (PFN) and have been implemented for both classification [60] and regression [45] tasks. Finally, it is interesting to note that *Large Language Models* (LLMs) have been successfully applied to both Zero-Shot classification [61] and regression [62] tasks without additional fine-tuning.

One-Shot Learning techniques generally rely on learning, as Meta-Representation, a similarity metric that can later on be used to classify unseen datapoints by comparing them with one example per target-dataset class. Examples of such architectures are *siamese networks* [63], and analogous models coming from different domains [64], and *matching networks* [65] (and similar variants [66]). Alternatively, there have also been successful applications of *Memory Augmented Neural Networks* [67], which are inherently great multi-task Meta-Learning architectures [68], to the One-shot Learning problem [69]. Such an application is made possible by the meta-information stored in the memory modules, which, in this case, represents the actual Meta-Representation of the algorithm. Other models that have excelled in OSL scenario are the *prototypical networks* [18], which build an embedding module (Meta-Representation) capable of producing a class prototype in a one-shot manner [70]; new examples are embedded and then compared for similarities with the various prototypes to determine their class.

The most widespread Meta-Representation in K-shot learning is parameter initialization as the techniques belonging to this scenario generally rely on fine-tuning for adapting a generic pre-trained model to a precise task. The most famous algorithms in this realm are MAML [38] and its variants [39, 40], but similar examples are coming also from domain-specific implementations [19, 71]. Some alternative frameworks that are based on fine-tuning are generating easily adaptable parameters on the fly based on a few examples of the target dataset [24] and Meta-Learning optimizers [30]. In a completely different direction, the authors of [72] show that it is possible to use a *Wasserstein Generative Adversarial Network* to generate synthetic training data from a few real examples; such implementation aims to obtain a data augmentation module as Meta-Representation.

3.3   Meta-Learning Example Formulation of ZSL, OSL and FSL for TSA and TSF

When we defined the various forms of N-shot learning, our unit of measure to determine within which definition each technique was falling under was the concept of "example". As a consequence, it is crucial to clearly state what what we mean by example in the case of time series data before further proceeding with our discussion.

As time series data lies within a structured domain endowed with an accessibility relation (the intuitive concept of "successor" in a series), it does not make sense to treat a single datapoint within a sequence as an example. Even if we worked with n-variate series in which each value belongs to $\mathbb{R}^n$, there is no temporal information to infer from a single point as seasonality and trend are concepts that can only be defined over ordered sets of values. Instead, it makes sense to treat as an example any sequence of size less or equal to the maximum history length seen during training. This provides a clear criterion to unambiguously locate any N-shot learning approach within the ZSL/OSL/FSL taxonomy, but without imposing any pre-determined limit to the size of a single example.

After clearly stating our definition of example in the case of time series data, the formulations of the three types of N-shot learning follow from the definitions provided in the previous subsection. In particular, we are now able to explain why time series data is one of the few domains where the definition of One-Shot Regression can make sense. Normally, it is impossible to fit a function from a single input-output mapping: think of interpolating a function $f : \mathbb{R} \to \mathbb{R}$ from a single $(x, f(x))$ pair: as there are infinite curves passing through it, there is no way to unambiguously determine the nature of f (unless there is some meta-data available, for example we may know that f is a linear function with a certain slope). On the contrary, in the case of time series data, x is a sequence of values, and, as a consequence, we can potentially train our predictor on all of its ordered subsequences to (ideally) extract seasonal patterns and trends from few datapoints. Such approach, although it clearly introduces various tedious challenges, paves the way for the implementation of OSL frameworks for forecasting.

After unambiguously defining the concept of example, the problem formulation of ZSL, OSL and FSL for time series data can be trivially derived from Sect. 3.2.

3.4   Pipeline Discussion for ZSL, OSL and FSL for TSF

Although the three N-shot learning approaches seem similar at first glance, their use of the few examples at disposal vastly varies from technique to technique; furthermore, within the same category different architectures may exploit data in completely different ways, so there is no "mandatory" pipeline used in this field (as it would essentially limit the creativity of the researchers). Nevertheless, within each approach we can outline a "generic" structure in the processing steps of the available examples and, therefore, we will now present the most common pipelines of each category of the taxonomy.

ZSL Pipeline. In the case of ZSL for Time Series Forecasting, a pipeline should focus on generalizing the temporal features of the source dataset. Such pipeline may thus include a combination of the following steps (see Fig. 2):

1. Feature Extraction and Representation: this step consists in extracting robust features that capture underlying patterns in examples coming from the source time series dataset.

2. Semantic Space Mapping: the extracted features are mapped into a latent semantic space where similarities between seen and unseen series/classes can be identified.
3. Forecast Model Training: exploiting the learned mappings, the predictor is meta-trained, while ensuring that some form of generalization mechanism is in place to maintain the performance after the shift from the source to the target dataset.
4. Prediction and Adaptation: after meta-training, it may be useful to expose the model to unseen time series and, in case, to refine it based on its validation performance.

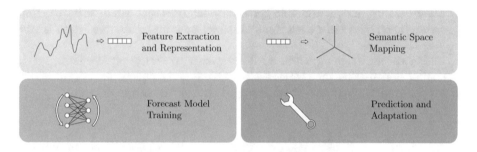

Fig. 2. Most common steps of ZSL Pipelines

OSL Pipeline. When developing OSL techniques, we have a single target example, or one example per class in the case of classification tasks, from which we can extract information with a pretrained model. Pipelines implemented in this case generally include (see Fig. 3):

1. Instance Selection: select one representative instance from each category of time series.
2. Feature Learning: learn from few instances some deep features that provide great generalization capabilities.
3. Similarity Modeling: learn a similarity metric that will help to compare new instances against the selected instances.

FSL Pipeline. FSL is the most general N-Shot Learning approach and, as such, it may involve multiple steps and techniques. Nevertheless, most FSL pipelines include a combination of (see Fig. 4):

1. Instance Subset Selection: for each time series category, select a very small number of representative instances.

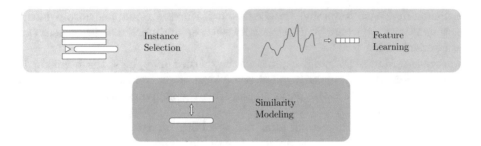

Fig. 3. Most common steps of OSL Pipelines

2. Meta-Features Extraction: find good meta-features that can describe the time series/time series class even if only a small amount of observed data is provided.
3. Meta-Learner Training: train a Meta-Learner to ensure rapid adaptation to similar, but unseen time series tasks.
4. Model Fine-Tuning: fine-tuned the predictor on the target dataset.

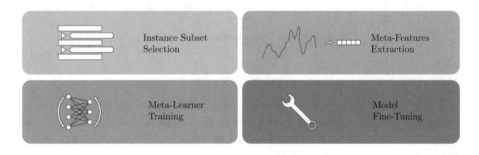

Fig. 4. Most common steps of FSL Pipelines

3.5 Comprehensive Review of Meta-Learning with Regard to ZSL, OSL and FSL for TSA/TSF

All the variety that comes with the breadth of definition of Meta-Learning is reflected in the time series domain by the sheer number of different approaches that have been presented in the past few years. For both classification and regression tasks, there have been multiple proposals of architectures and algorithms that leverage previous knowledge through different strategies for an improvement in efficiency and/or performance. In this section, we will provide a review of the recent results of M-L for time series data, showing also where they fit within the previously introduced [3] taxonomy.

A possible route for exploiting previous Meta-Learning knowledge in Time Series Forecasting could be model selection (Meta-Objective). This concept has been proposed as *Feature-Based Forecast Model Selection* (FFORMS) [73] and aims to learn, as Meta-Representation, the mapping between time series features and the suggested pre-trained architectures. Such an idea has been later validated in [74], where the authors additionally explore the possibility of combining the forecasts of multiple models instead of relying on the expressiveness of a single architecture. Similarly, in [75], the authors propose a Meta-Learning-based recommendation system for time series classifiers based on statistics computed on individual series. These methods belong to the 0-shot learning realm, as they do not require any further fine-tuning on the target series. Other successful Zero-Shot Learning approaches for forecasting have considered either pre-training a model on real [76], or synthetic [45] data, or exploiting LLMs' natural tendency towards token prediction in sequences [62].

If, instead, we have access to at least a restricted target dataset for fine-tuning, the Time Series Forecasting problem can also be tackled using model-agnostic techniques [7,19]. Such frameworks exploit meta-training over multiple episodes (gradient descent as Meta-Optimizer) to obtain either an easily fine-tunable set of initial parameters (Meta-Representation) for an unseen task, or a similarity measure to efficiently compute a prediction by comparing a target history with similar examples from the support set [77].

Classification problems, on the other hand, generally benefit from Meta-Learning how to efficiently create a representation (Meta-Representation) of the classes in a given dataset. This could be done by learning task-agnostic embedding modules [56] or prototypical networks [48] that can provide respectively Zero-Shot and One-Shot Learning capabilities. In contrast, in the Few-Shot Learning realm, it has been shown that residual connections in Neural Networks can act as Meta-Learning adaptation mechanisms by providing a set of task-specific parameters [6]. This idea has been successfully implemented in the classification domain [78] by meta-training a *ResNet* on a heterogeneous set of Time Series Classification problems reformulated in terms of a task-agnostic loss [79] and then fine-tuning it on the target distribution.

4   Illustrative Implementations of ZSL, OSL and FSL for TSF (RQ3)

As part of the contributions of this survey, we aimed to study the actual benefits gained by applying different Meta-Learning methods to the same problem to highlight the pros and cons of each approach. Our study will show how the implementation of 3 different M-L techniques (ZSL, OSL and FSL) will alter the performance of a base ML learner on the same data. In particular, for better comparability, we decided to keep the base architecture unaltered and to apply the least amount of modifications required for each approach.

Note that, to improve clarity, in this section we included a set of figures to visualize how data is used during training/testing by the various algorithms. In such figures, the part of the dataset employed by a certain model is highlight in black, whereas the examples that are left out are drawn in light gray.

4.1   Dataset Presentation

To ensure a fair comparison, it was necessary to find a dataset with sufficient data and variety to allow for M-L adaptations to manifest: we chose the *Yahoo Finance Dataset* available at [80], which holds the histories of almost 500 stocks. Apart from 20 of them, which we left out of our analysis, all the companies' histories have more than 1000 values distributed between the 29/11/2018 and 29/11/2023. The first step towards a meaningful comparison is to decide how to split the data into source and target datasets. Unambiguously defining such partitioning is the foundation of any rigorous test within this field, as it allows to clearly identify which data we can use to meta-train/fine-tune/test our models. Since all the companies left from the initial filtering have the same amount of datapoints, we kept $\frac{2}{3}$ of the stocks, choosen at random, as meta-training dataset $\mathcal{D}_{\text{source}}$. The remaining third, $\mathcal{D}_{\text{target}}$ is furtherly split into a fine-tuning dataset $\mathcal{D}_{\text{f-t}}$ and a testing dataset $\mathcal{D}_{\text{test}}$ (in practice, $\mathcal{D}_{\text{f-t}} \cup \mathcal{D}_{\text{test}} = \mathcal{D}_{\text{target}}$) based on the average timestamp (the second of June 2021 according to our analysis). This partitioning was done to strike a balance between a "big-enough" test set to provide relevant results and a "bigger-enough" source dataset to implement M-L techniques correctly.

4.2   Setup of the Implementation

For further clarity, we will define the set of all the companies under analysis as $\mathcal{C}$, the subsets of source and target companies as $\mathcal{C}_{\text{source}}, \mathcal{C}_{\text{target}} \subseteq \mathcal{C}$, with $\mathcal{C}_{\text{source}} \cap \mathcal{C}_{\text{target}} = \emptyset$ and the data of a dataset $\mathcal{D}$ belonging to a particular company c as $\mathcal{D}^{(c)}$. This partitioning approach is displayed in Fig. 5.

The architecture implemented is a standard two-layer encoder [81] with a maximum sequence size of 100, 8 attention heads, and an internal representation size of 32^{1}. The data is fed to the transformer as a 100×5 tensor, where each column represents a single timestamp of the history and the rows contain:

1. The year of the timestamp, min-max scaled.
2. The month of the timestamp, min-max scaled.
3. The day of the timestamp, min-max scaled.
4. The day of the week of the timestamp, min-max scaled.
5. The value of the time series, robust scaled (for more information about robust scaling, we refer to [45]).

Note that, following the choices of [45], all the values are scaled between 0 and 3, and that before example extraction time series are split into 100 datapoints-long subseries and scaled individually. From a single subseries, examples are

---

[1] These hyperparameters were derived from a tuning procedure of the ZSL method.

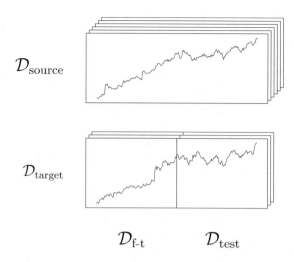

$\mathcal{D}_{\text{source}}$

$\mathcal{D}_{\text{target}}$

$\mathcal{D}_{\text{f-t}}$ $\mathcal{D}_{\text{test}}$

Fig. 5. Partitioning of the data for the experiment

extracted sequentially, with incrementally longer histories ranging from a length of 6 all the way to 97. The input sequence of the encoder is composed by a *start of sequence* symbol, the actual history, a *end of sequence* symbol and a *padding matrix* of variable non-zero length, thus the reason for which the maximum sequence size is 97. Along with the input, the encoder is also fed a boolean mask that distinguishes the actual input from the padding. The output of the architecture is a single numerical result, the next (scaled) value of the time series.

4.3 Discussion of Performance Metrics of Relevance to ZSL/OSL/FSL

Normally, the performance of Machine Learning models is evaluated only through accuracy and error rates, where accuracy is defined as the proportion of correct predictions in classification tasks and error rates quantify in regression contexts metric values such as Mean Absolute Error (MAE) and Root Mean Squared Error (RMSE). In particular, in the case of N-Shot Learning, such metrics are meaningful with regards to examples coming from unseen distributions.

Apart from the standard ML evaluation metrics, the case of M-L another salient dimension to be measured are the generalization capabilities of the model. This consists in evaluating how good a predictor that is trained under the ZSL, OSL or FSL regime performs when exposed to completely new data. Good generalization capabilities are actually a testament to the efficacy of a model in real-world deployments, where data will (possibly) always be varying. Robustness to noise and incomplete data may also be key requirements for a M-L model that may be tested before production. Generalization capabilities can be tested by exposing the model to data coming from distributions progressively more distant from the source dataset and more complex. In the case of FSL, more specific metrics such as the learning curves [56] over few-shot scenarios and improvement rates with additional data points may also provide more detailed insights into the models' Meta-Learning performances.

Convergence speed is one of the key factors in M-L, especially for FSL where models should be ready to rapidly adapt to new data: the faster a model can reliably learn from few examples, the more efficient it is deemed in a FSL scenario. Metrics that can evaluate the computational complexity of deploying a certain model can be obtained as simply as measuring the time required for meta-training and/or providing data about its speed of convergence, such as the standard training-validation error graphs, along with a detailed description of the hardware used for the process.

In sum, these performance metrics help not only in relative comparison of different models but also bring out the practical considerations and challenges involved in implementing these models in real world M-L tasks, as they provide a holistic view of a model reliability, adaptability, efficiency, and scalability in diverse scenarios.

4.4   ML Baseline Implementation: Standard Encoder

Our base model is trained directly on the fine-tuning dataset $\mathcal{D}_{\text{f-t}}$, on a "company-basis". For each stock that belongs to the target set $c \in \mathcal{C}_{\text{target}}$, we train the architecture from scratch on the first half of its datapoints $\mathcal{D}_{f-t}^{(c)}$ and then test it on the second half $\mathcal{D}_{test}^{(c)}$. This can be considered as a standard ML approach as we are using only a limited amount of data of a single stock to predict its next values, without exploiting external knowledge. The dataset partitioning approach is shown in Fig. 6.

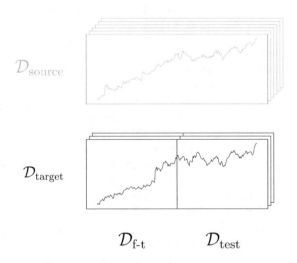

Fig. 6. Data used for training and testing of the ML model

4.5   ZSL Forecasting Implementation: ForecastPFN

The ZSL approach is heavily inspired by the *ForecastPFN* framework introduced in [45]: the model is trained using entirely synthetic data, procedurally generated in 10 batches of 100 series of length 200, subsequently splitted into 100-datapoints-long subsequences using a sliding window approach. For more information about the generation of the data, readers can refer to [45]. In the ZSL case, we disregard completely the source dataset $\mathcal{D}_{\text{source}}$, and test directly onto the target dataset $\mathcal{D}_{\text{test}}$ (see Fig. 7).

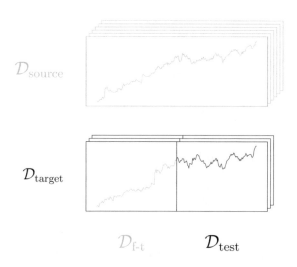

Fig. 7. Data used for training and testing of the ZSL model

4.6   OSL Forecasting Implementation: Siamese Networks

Following the idea of [77], we introduce a OSL predictor that exploits a single subsequence of a company's history to forecast the subsequent values of its stock. This architecture uses a *Siamese Neural Network* [63] to learn a difference function between two histories: given hist_{ref} and $\text{hist}_{\text{test}}$, it passes them through the same network $\mathcal{N}$ to produce $\text{forecast}_{\text{ref}} = \mathcal{N}(\text{hist}_{\text{ref}})$ and $\text{forecast}_{\text{test}} = \mathcal{N}(\text{hist}_{\text{test}})$, and output the value $\text{forecast}_{\text{ref}} - \text{forecast}_{\text{test}}$. This allows us to meta-train the network on all pairs of subsequences derived from each company's history in the source dataset $\mathcal{D}_{\text{source}}$ and test the architecture directly on each stock $c \in \mathcal{C}_{\text{target}}$ by taking the first 100 datapoints of $\mathcal{D}_{\text{f-t}}^{(c)}$ as hist_{ref} and all subsequence in $\mathcal{D}_{\text{test}}^{(c)}$ as $\text{hist}_{\text{test}}$. The dataset partitioning approach is shown in Fig. 8.

In the original paper [77], the authors used a CNN-LSTM network for realizing $\mathcal{N}$. But to keep our comparison relevant, we implement $\mathcal{N}$ using the same encoder architecture previously discussed.

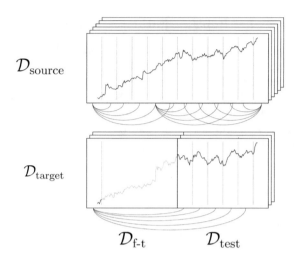

Fig. 8. Data used for training and testing of the OSL model

4.7   FSL Forecasting Implementation: Reptile

For our FSL approach we decided to adopt a MAML approximation: the REP-TILE framework [40] by OPENAI. In this case, the architecture, which is the same as in the previous sections, gets trained with all the data contained in $\mathcal{D}_{\text{source}}$ and then, for each company $c \in \mathcal{C}_{\text{target}}$, it is fine-tuned on $\mathcal{D}_{\text{f-t}}^{(c)}$, before being tested on $\mathcal{D}_{\text{test}}^{(c)}$ (see Fig. 9). This approach provides additional meta-knowledge to the learner, which, hopefully, gets to learn a good parameter initialization for quick fine tuning during meta-training.

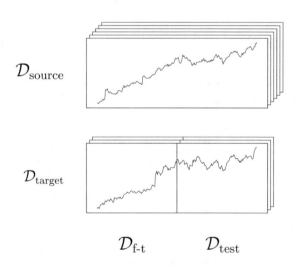

Fig. 9. Data used for training and testing of the FSL model

4.8   Comparative Discussion of the Results

Our metric of choice for the comparison is the Root Mean Square Error (RMSE). A comprehensive table of results is available in Table 1; however, for convenience, we randomly sampled a limited set of stocks and compared the performance metrics of the various models on them. As we can see in Fig. 10, the results are pretty consistent across the examples. The additional meta-knowledge acquired by the MAML-based method led to a marginal improvement in the performance of the FSL model, whereas the other M-L approaches have shown drastically worse results, even compared to the base ML predictor. This decrease in performance is worth to be addressed individually; the OSL model has originally been created to solve tasks with time series that have a really strong periodic component: by aligning the periods of two inputs coming from the same distribution, the siamese network is able, at least in the original article, to successfully predict new values based on the reference history. In the case of stock predictions, the examples do not generally manifest a strongly periodic behaviour and instead are heavily dominated by the trend and noise components, making this model's forecast unreliable. The case of ZSL is different: to have good generalization capabilities, the model must be trained extensively on a huge set of generated time series; otherwise, it will not have the generalization capabilities for tackling a forecasting task on such volatile data as stocks. We may not have trained the model long enough to see the real capabilities of the Meta-Learning adaptations. Synthetic data allows us also to use predictors of biggers sizes, as we can avoid the risk of underfitting by simply ensuring that the pseudorandom generation of data is general enough. We leave as future work to analyze the asymptotical performance of increasingly more complex predictors in tandem with increasingly longer training regimes.

5   Challenges and Limitations of Current Meta-Learning Techniques and Directions for Future Research in the Context of TSF/TSA

The incorporation of Meta-Learning as a new frontier in Machine Learning introduces new, exciting opportunities for enhancing Time Series Forecasting (TSF) and Time Series Analysis (TSA). Despite their promising performance, the existing M-L approaches present a mix of challenges and limitations, primarily brought by the inherent complexities of time series data. In this section, we will highlight the main challenges related to this field, along with some outlook of the possible future directions of research.

5.1   Challenges and Limitations of Meta-Learning for TSA/TSF

While showcasing new horizons in Time Series Analysis and Forecasting, Meta-Learning faces several challenges and limitations. Identifying the main recurrent challenges can help in focusing on finding the solutions to the pertinent problems that may prevent Meta-Learning from being an effective tool in real-world applications.

Table 1. RMSE values for the Baseline, FSL, OSL and ZSL models

Company	ML Baseline	FSL	OSL	ZSL
ACGL	0.09688	0.09773	0.95851	0.62487
ADP	0.08762	0.08556	0.95934	0.62214
AEM	0.14746	0.12432	0.98937	0.59860
AMZN	0.12812	0.13390	0.97672	0.60874
AZO	0.07321	0.07118	0.95231	0.63269
BBD	0.11465	0.11554	0.96578	0.62379
BHP	0.11712	0.11895	0.96314	0.60490
BIIB	0.11417	0.11148	0.91524	0.60663
BKR	0.10457	0.11140	0.95046	0.62283
BMO	0.09104	0.08593	0.97308	0.61453
BMY	0.07202	0.07863	0.98159	0.61336
BN	0.09588	0.09529	0.96636	0.62141
CAH	0.08187	0.08301	0.95763	0.61163
COR	0.07097	0.06574	0.96823	0.62550
CTAS	0.09196	0.11937	0.95780	0.60759
CVS	0.07808	0.07343	0.92659	0.60973
DHI	0.11621	0.12409	0.98228	0.60345
DXCM	0.18518	0.17374	0.95139	0.63579
EL	0.12195	0.12678	0.94381	0.60244
EW	0.10873	0.11576	0.96374	0.59995
FERG	0.10679	0.10475	0.99167	0.60550
GD	0.06645	0.07691	0.98904	0.62433
GILD	0.08941	0.08909	0.99410	0.61996
GM	0.12947	0.13111	0.98307	0.60784
GMAB	0.11411	0.11182	0.97473	0.60902
HCA	0.11479	0.10001	0.96931	0.61632
KHC	0.10253	0.09127	0.88149	0.60160
KR	0.09185	0.08221	0.96799	0.61282
LOW	0.10692	0.12192	0.96582	0.61225
MPWR	0.15202	0.14978	0.97049	0.62230
MSI	0.08731	0.08420	0.95215	0.59721
NOK	0.10679	0.08983	0.99015	0.61147
NUE	0.11722	0.10652	0.98447	0.62356
NVS	0.08196	0.07193	0.97910	0.61415
NWG	0.08006	0.08645	0.95151	0.60835
OXY	0.18388	0.17538	1.00188	0.61009
PCAR	0.07570	0.07057	0.95987	0.62122
SAN	0.10491	0.10725	0.99158	0.59907
SBAC	0.10135	0.10626	0.97216	0.60495
SMFG	0.08324	0.08021	0.99504	0.61343
SNY	0.08276	0.07004	0.99733	0.60720
SPOT	0.17818	0.17517	0.97748	0.62467
STM	0.11416	0.12489	0.95955	0.61805
T	0.07902	0.07600	0.98919	0.59949
TDG	0.09544	0.10842	0.93145	0.61190
TEL	0.09401	0.10378	0.97777	0.61068
TTE	0.10290	0.09929	0.99839	0.61080
UNH	0.07448	0.06848	0.97275	0.61793
WDAY	0.14312	0.13112	0.96546	0.62501
WDS	0.09467	0.10321	0.98014	0.60773

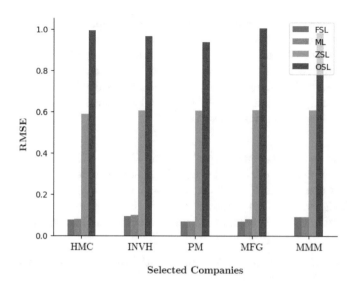

Fig. 10. RMSE performance for the ML Baseline, ZSL, OSL and FSL models of randomly selected companies

Heterogeneity and Non-stationarity of Data. One of the main challenges is the non-stationarity of time series data inherent to some domains: in such cases, the statistical properties of the data vary as its patterns change over time, making it hard for models to learn an effective generalization. Even though ensemble methods have shown promising results in handling such complexity [82], this characteristic of time series Data remains an issue that requires careful evaluation.

Data Limitation and Overfitting. Datasets of limited size may lead to underfitting, resulting in models featuring poor generalization capabilities for unseen data. A potentially intuitive (non-)solution may be to iteratively train predictors on a restricted set of datapoints that may not be representative of the actual target distribution, causing overfitting on the source data. The challenge is to develop M-L algorithms that can learn from small sample sizes but without ending up in one of the pitfalls described above.

Complexity of Model Selection and Hyperparameter Tuning. When dealing with M-L, introducing an extra layer of learning (learning to learn), the model selection and hyperparameters-tuning become even harder. In the case of Time Series Forecasting such increase in complexity may even be amplified due to the existence of time dependencies and possibilities of multiple seasonal patterns.

Scalability and Computational Resources. Application of Meta-Learning techniques to Deep Learning models is generally resource-intensive, as the second layer of optimization can potentially make the training process more computationally expensive. Even though some approaches (for example those that involve pretraining and fine-tuning phases) may actually lead to amortized costs when deploying multiple models, scalability must be taken into consideration when dealing with large scale time series data.

Generalizes to Different Time Series Tasks. The objective of Meta-Learning is to create models that will quickly adapt to new distribution characteristics, but as each TSF/TSA task tends to have unique features and patterns, it's not straightforward for a model which was meta-learned to generalize well across the different time series tasks with no or minimal re-training.

5.2 Suggesting Potential Avenues and Directions for Future Research in TSA/TSF

Meta-Learning has been successfully implemented in multiple domains, time series data included. Yet, the challenges presented in the previous section highlight that there are some directions that need to be further investigated. In the remaining part of this section we will discuss some of the most promising avenues of research in this field.

Hybrid Models and Feature Engineering. Traditional statistical techniques have been successfully implemented for time series tasks for decades, featuring a much longer history of applications than Machine Learning, let alone Deep Learning. Although ML methods have recently shown very promising results, traditional methods feature some unmatched advantages, such as explainability and lightweight implementations. Hybrid models featuring M-L techniques may integrate the strengths of both approaches, leading to more accurate and accessible models. Additionally, advanced feature engineering [56] may enhance the capabilities of the predictors by extracting meaningful statistics and meta-data that could compensate for the complexity of time series data.

Meta-Learning in Multivariate and High-Dimensional Time Series. Most of the current TSA/TSF techniques are oriented towards univariate time series [19,62]. Integrating the information coming from multiple sources may provide additional contextual data that M-L models could exploit for learning better generalizations, but this clearly requires designing innovative methods that help extract and actually use the combined information. Future research towards this objective may provide noticeable improvements to the analysis and forecasting of stream-like data in various domains, such as healthcare, where multiple sensors may be used at once, and financial forecasting, where the influence of the external world plays a clear role in the fluctuations of stocks.

Representation Learning with Meta-Learning. Another possible direction of research is to investigate how representation learning [8] can be added to Meta-Learning frameworks to further improve TSF and TSA. By learning rich and meaningful representations that are able to capture the underlying dynamic structure of time series data we may be able to produce more accurate predictions, maybe also requiring less computationally-intensive training procedures.

Meta-Learning for Real-Time Forecasting. Another significant line of research is the creation of real-time forecasting Meta-Learning models for dynamic environments such as financial markets or IoT applications. Such applications require an advanced methodology to ensure prompt adaptation to new data as soon as it becomes available.

6 Conclusion

This work has addressed the application of Meta-Learning in Time Series Forecasting and Analysis. Achieved by presenting a consolidated discussion that comprehensively contrasts ZSL, OSL, and FSL, the results of the various implementations have been presented. Implemented as a MAML-based method, FSL presents the best performance due to the additional generalization capabilities obtained through meta-training. On the other hand, OSL, implemented as a Siamese Network, shows the worse performance, probably caused by the lack of strong periodicity in the data. Next, the work highlighted some challenges and limitations faced in Meta-Learning for Time Series Analysis and Forecasting, among which we can identify heterogeneity and non-stationarity of data, overfitting, complexity of model selection and hyperparameter tuning, and generalization to different time series task distributions. Finally, the work discusses some potential avenues of future research that could advance Meta-Learning in TSA/TSF.

References

1. Song, Y., Wang, T., Cai, P., Mondal, S., Sahoo, J.: A comprehensive survey of few-shot learning: evolution, applications, challenges, and opportunities. ACM Comput. Surv. **55**, 1–40 (2023)
2. Shi, J., Jain, M., Narasimhan, G.: Time series forecasting (TSF) using various deep learning models. arXiv Preprint arXiv:2204.11115 (2022)
3. Hospedales, T., Antoniou, A., Micaelli, P., Storkey, A.: Meta-learning in neural networks: a survey. IEEE Trans. Pattern Anal. Mach. Intell. **44**, 5149–5169 (2021)
4. Tian, Y., Zhao, X., Huang, W.: Meta-learning approaches for learning-to-learn in deep learning: a survey. Neurocomputing **494**, 203–223 (2022)
5. Wang, Y., Ramanan, D., Hebert, M.: Meta-learning to detect rare objects. In: Proceedings of the IEEE/CVF International Conference on Computer Vision, pp. 9925–9934 (2019)

6. Oreshkin, B., Carpov, D., Chapados, N., Bengio, Y.: Meta-learning framework with applications to zero-shot time-series forecasting. In: Proceedings of the AAAI Conference on Artificial Intelligence, vol. 35, pp. 9242–9250 (2021)
7. Xiao, F., et al.: Meta-learning for few-shot time series forecasting. J. Intell. Fuzzy Syst. **43**, 325–341 (2022)
8. Zhang, J., Ghahramani, Z., Yang, Y.: Flexible latent variable models for multi-task learning. Mach. Learn. **73**, 221–242 (2008)
9. Zhang, Y., Wei, B., Li, X., Li, L.: A survey of meta-learning for classification tasks. In: 2022 10th International Conference on Information Systems and Computing Technology (ISCTech), pp. 442–449 (2022)
10. Ahmed, M., Mahmood, A., Hu, J.: A survey of network anomaly detection techniques. J. Netw. Comput. Appl. **60**, 19–31 (2016)
11. Bichri, H., Chergui, A., Hain, M.: Image classification with transfer learning using a custom dataset: comparative study. Procedia Comput. Sci. **220**, 48–54 (2023)
12. Kim, H., Cosa-Linan, A., Santhanam, N., Jannesari, M., Maros, M., Ganslandt, T.: Transfer learning for medical image classification: a literature review. BMC Med. Imaging **22**, 69 (2022)
13. Xu, G., et al.: A deep transfer convolutional neural network framework for EEG signal classification. IEEE Access **7**, 112767–112776 (2019)
14. Subramanian, M., Sathishkumar, V., Cho, J., Shanmugavadivel, K.: Learning without forgetting by leveraging transfer learning for detecting COVID-19 infection from CT images. Sci. Rep. **13**, 8516 (2023)
15. Iman, M., Arabnia, H., Rasheed, K.: A review of deep transfer learning and recent advancements. Technologies **11**, 40 (2023)
16. Lee, K., Jung, S., Ryu, J., Shin, S., Choi, J.: Evaluation of transfer learning with deep convolutional neural networks for screening osteoporosis in dental panoramic radiographs. J. Clin. Med. **9**, 392 (2020)
17. Hosna, A., Merry, E., Gyalmo, J., Alom, Z., Aung, Z., Azim, M.: Transfer learning: a friendly introduction. J. Big Data **9**, 102 (2022)
18. Snell, J., Swersky, K., Zemel, R.: Prototypical networks for few-shot learning. In: Advances in Neural Information Processing Systems, vol. 30 (2017)
19. Iwata, T., Kumagai, A.: Few-shot learning for time-series forecasting. arXiv Preprint arXiv:2009.14379 (2020)
20. Gogoi, M., Tiwari, S., Verma, S.: Adaptive prototypical networks. arXiv Preprint arXiv:2211.12479 (2022)
21. Amasyali, M., Ersoy, O.: A study of meta learning for regression. ECE Technical Reports, p. 386 (2009)
22. Bengio, Y.: Gradient-based optimization of hyperparameters. Neural Comput. **12**, 1889–1900 (2000)
23. Rajeswaran, A., Finn, C., Kakade, S., Levine, S.: Meta-learning with implicit gradients. In: Advances in Neural Information Processing Systems, vol. 32 (2019)
24. Munkhdalai, T., Yu, H.: Meta networks. In: International Conference on Machine Learning, pp. 2554–2563 (2017)
25. Guo, N., Di, K., Liu, H., Wang, Y., Qiao, J.: A metric-based meta-learning approach combined attention mechanism and ensemble learning for few-shot learning. Displays **70**, 102065 (2021)
26. Gao, B., Gouk, H., Yang, Y., Hospedales, T.: Loss function learning for domain generalization by implicit gradient. In: International Conference on Machine Learning, pp. 7002–7016 (2022)
27. Chauhan, V., Zhou, J., Lu, P., Molaei, S., Clifton, D.: A brief review of hypernetworks in deep learning. arXiv Preprint arXiv:2306.06955 (2023)

28. Finn, C., Levine, S.: Meta-learning and universality: deep representations and gradient descent can approximate any learning algorithm. arXiv Preprint arXiv:1710.11622 (2017)
29. Andrychowicz, M., et al.: Learning to learn by gradient descent by gradient descent. In: Advances in Neural Information Processing Systems, vol. 29 (2016)
30. Li, Z., Zhou, F., Chen, F., Li, H.: Meta-SGD: learning to learn quickly for few-shot learning. arXiv Preprint arXiv:1707.09835 (2017)
31. Liu, Y., Sun, Y., Xue, B., Zhang, M., Yen, G., Tan, K.: A survey on evolutionary neural architecture search. IEEE Trans. Neural Netw. Learn. Syst. **34**, 550–570 (2021)
32. Alet, F., Weng, E., Lozano-Pérez, T., Kaelbling, L.: Neural relational inference with fast modular meta-learning. In: Advances in Neural Information Processing Systems, vol. 32 (2019)
33. Wang, Y., Yao, Q., Kwok, J., Ni, L.: Generalizing from a few examples: a survey on few-shot learning. ACM Comput. Surv. (CSUR) **53**, 1–34 (2020)
34. Xiang, H., Lin, J., Chen, C., Kong, Y.: Asymptotic meta learning for cross validation of models for financial data. IEEE Intell. Syst. **35**, 16–24 (2020)
35. Schmidhuber, J., Zhao, J., Wiering, M.: Shifting inductive bias with success-story algorithm, adaptive Levin search, and incremental self-improvement. Mach. Learn. **28**, 105–130 (1997)
36. Zhang, Y., Yang, Q.: A survey on multi-task learning. IEEE Trans. Knowl. Data Eng. **34**, 5586–5609 (2021)
37. Finn, C., Rajeswaran, A., Kakade, S., Levine, S.: Online meta-learning. In: International Conference on Machine Learning, pp. 1920–1930 (2019)
38. Finn, C., Abbeel, P., Levine, S.: Model-agnostic meta-learning for fast adaptation of deep networks. In: International Conference on Machine Learning, pp. 1126–1135 (2017)
39. Fan, C., Ram, P., Liu, S.: Sign-MAML: efficient model-agnostic meta-learning by SignSGD. arXiv Preprint arXiv:2109.07497 (2021)
40. Nichol, A., Achiam, J., Schulman, J.: On first-order meta-learning algorithms. arXiv Preprint arXiv:1803.02999 (2018)
41. Gonzalez, S., Miikkulainen, R.: Improved training speed, accuracy, and data utilization through loss function optimization. In: 2020 IEEE Congress on Evolutionary Computation (CEC), pp. 1–8 (2020)
42. Cubuk, E., Zoph, B., Mane, D., Vasudevan, V., Le, Q.: Autoaugment: learning augmentation policies from data. arXiv Preprint arXiv:1805.09501 (2018)
43. Gupta, A., Raghav, Y.: Time Series Classification with Meta Learning. AIRCC Publishing Corporation (2020)
44. Brinkmeyer, L., Drumond, R.R., Burchert, J., Schmidt-Thieme, L.: Few-shot forecasting of time-series with heterogeneous channels. In: Amini, M.R., Canu, S., Fischer, A., Guns, T., Kralj Novak, P., Tsoumakas, G. (eds.) ECML PKDD 2022. LNCS, vol. 13718, pp. 3–18. Springer, Cham (2023). https://doi.org/10.1007/978-3-031-26422-1_1
45. Dooley, S., Khurana, G., Mohapatra, C., Naidu, S., White, C.: ForecastPFN: synthetically-trained zero-shot forecasting. arXiv Preprint arXiv:2311.01933 (2023)
46. Jiang, R., et al.: Spatio-temporal meta-graph learning for traffic forecasting. In: Proceedings of the AAAI Conference on Artificial Intelligence, vol. 37, pp. 8078–8086 (2023)
47. Oreshkin, B., Carpov, D., Chapados, N., Bengio, Y.: N-BEATS: neural basis expansion analysis for interpretable time series forecasting. arXiv Preprint arXiv:1905.10437 (2019)

48. Tang, W., Liu, L., Long, G.: Interpretable time-series classification on few-shot samples. In: 2020 International Joint Conference on Neural Networks (IJCNN), pp. 1–8 (2020)
49. Smyl, S.: A hybrid method of exponential smoothing and recurrent neural networks for time series forecasting. Int. J. Forecast. **36**, 75–85 (2020)
50. Montero-Manso, P., Athanasopoulos, G., Hyndman, R., Talagala, T.: FFORMA: feature-based forecast model averaging. Int. J. Forecast. **36**, 86–92 (2020)
51. Zyl, T.: Late meta-learning fusion using representation learning for time series forecasting. arXiv Preprint arXiv:2303.11000 (2023)
52. Pourpanah, F., et al.: A review of generalized zero-shot learning methods. IEEE Trans. Pattern Anal. Mach. Intell. **45**, 4051–4070 (2022)
53. Chen, L., Zhang, H., Xiao, J., Liu, W., Chang, S.: Zero-shot visual recognition using semantics-preserving adversarial embedding networks. In: Proceedings of the IEEE Conference on Computer Vision and Pattern Recognition, pp. 1043–1052 (2018)
54. Shigeto, Y., Suzuki, I., Hara, K., Shimbo, M., Matsumoto, Y.: Ridge regression, hubness, and zero-shot learning. In: Appice, A., Rodrigues, P.P., Santos Costa, V., Soares, C., Gama, J., Jorge, A. (eds.) ECML PKDD 2015. LNCS (LNAI), vol. 9284, pp. 135–151. Springer, Cham (2015). https://doi.org/10.1007/978-3-319-23528-8_9
55. Zhang, L., et al.: Towards effective deep embedding for zero-shot learning. IEEE Trans. Circuits Syst. Video Technol. **30**, 2843–2852 (2020)
56. Bhaskarpandit, S., Gupta, P., Gupta, M.: LETS-GZSL: a latent embedding model for time series generalized zero shot learning. arXiv Preprint arXiv:2207.12007 (2022)
57. Guo, C., Pleiss, G., Sun, Y., Weinberger, K.: On calibration of modern neural networks. In: International Conference on Machine Learning, pp. 1321–1330 (2017)
58. Chao, W.-L., Changpinyo, S., Gong, B., Sha, F.: An empirical study and analysis of generalized zero-shot learning for object recognition in the wild. In: Leibe, B., Matas, J., Sebe, N., Welling, M. (eds.) ECCV 2016. LNCS, vol. 9906, pp. 52–68. Springer, Cham (2016). https://doi.org/10.1007/978-3-319-46475-6_4
59. Bucher, M., Herbin, S., Jurie, F.: Generating visual representations for zero-shot classification. In: Proceedings of the IEEE International Conference on Computer Vision Workshops, pp. 2666–2673 (2017)
60. Hollmann, N., Müller, S., Eggensperger, K., Hutter, F.: TabPFN: a transformer that solves small tabular classification problems in a second. arXiv Preprint arXiv:2207.01848 (2022)
61. Chae, Y., Davidson, T.: Large language models for text classification: from zero-shot learning to fine-tuning. Open Science Foundation (2023)
62. Gruver, N., Finzi, M., Qiu, S., Wilson, A.: Large language models are zero-shot time series forecasters. arXiv Preprint arXiv:2310.07820 (2023)
63. Koch, G., Zemel, R., Salakhutdinov, R., et al.: Siamese neural networks for one-shot image recognition. In: ICML Deep Learning Workshop, vol. 2 (2015)
64. Zhuang, Z., Kong, X., Rundensteiner, E., Arora, A., Zouaoui, J.: One-shot learning on attributed sequences. In: 2018 IEEE International Conference on Big Data (Big Data), pp. 921–930 (2018)
65. Vinyals, O., Blundell, C., Lillicrap, T., Wierstra, D., et al.: Matching networks for one shot learning. In: Advances in Neural Information Processing Systems, vol. 29 (2016)
66. Altae-Tran, H., Ramsundar, B., Pappu, A., Pande, V.: Low data drug discovery with one-shot learning. ACS Cent. Sci. **3**, 283–293 (2017)
67. Graves, A., Wayne, G., Danihelka, I.: Neural turing machines. arXiv Preprint arXiv:1410.5401 (2014)

68. Graves, A., et al.: Hybrid computing using a neural network with dynamic external memory. Nature **538**, 471–476 (2016)
69. Santoro, A., Bartunov, S., Botvinick, M., Wierstra, D., Lillicrap, T.: Meta-learning with memory-augmented neural networks. In: International Conference on Machine Learning, pp. 1842–1850 (2016)
70. Boney, R., Ilin, A., et al.: Active one-shot learning with prototypical networks. In: ESANN (2019)
71. Bailer, W., Fassold, H.: Few-Shot Object Detection Using Online Random Forests
72. Li, K., Zhang, Y., Li, K., Fu, Y.: Adversarial feature hallucination networks for few-shot learning. In: Proceedings of the IEEE/CVF Conference on Computer Vision and Pattern Recognition, pp. 13470–13479 (2020)
73. Talagala, T., Hyndman, R., Athanasopoulos, G.: Meta-learning how to forecast time series. J. Forecast. **42**, 1476–1501 (2023)
74. Lemke, C., Gabrys, B.: Meta-learning for time series forecasting and forecast combination. Neurocomputing **73**, 2006–2016 (2010)
75. Abanda, A., et al.: Contributions to Time Series Classification: Meta-Learning and Explainability (2021)
76. Orozco, B., Roberts, S.: Zero-shot and few-shot time series forecasting with ordinal regression recurrent neural networks. arXiv Preprint arXiv:2003.12162 (2020)
77. Tran, V., Panangadan, A.: Few-shot time-series forecasting with application for vehicular traffic flow. In: 2022 IEEE 23rd International Conference on Information Reuse and Integration for Data Science (IRI), pp. 20–26 (2022)
78. Narwariya, J., Malhotra, P., Vig, L., Shroff, G., Vishnu, T.: Meta-learning for few-shot time series classification. In: Proceedings of the 7th ACM IKDD CoDS and 25th COMAD, pp. 28–36 (2020)
79. Schultz, M., Joachims, T.: Learning a distance metric from relative comparisons. In: Advances in Neural Information Processing Systems, vol. 16 (2003)
80. Arora, S.: Yahoo Finance Dataset (2023). figshare https://www.kaggle.com/datasets/suruchiarora/yahoo-finance-dataset-2018-2023
81. Vaswani, A., et al.: Attention is all you need. In: Advances in Neural Information Processing Systems, vol. 30 (2017)
82. Fazla, A., Aydin, M., Tamyigit, O., Kozat, S.: Context-aware ensemble learning for time series. arXiv Preprint arXiv:2211.16884 (2022)

Ensemble Learning
with Physics-Informed Neural Networks
for Harsh Time Series Analysis

Antoine Kazadi Kayisu[2], Paraskevi Fasouli[1],
Witesyavwirwa Vianney Kambale[1], Pitshou Bokoro[3],
and Kyandoghere Kyamakya[1,2(✉)]

[1] Institute for Smart Systems Technologies, Universität Klagenfurt,
Universitätsstraße 65/67, 9020 Klagenfurt, Austria
`kyandoghere.kyamakya@aau.at`
[2] Faculté Polytechnique, Université de Kinshasa,
Kinshasa, Democratic Republic of the Congo
`antoine.kayisu@unikin.ac.cd`
[3] Department of Electrical and Electronic Engineering Technology,
University of Johannesburg, Johannesburg, Republic of South Africa
`pitshoub@uj.ac.za`

Abstract. In time series data analysis, particularly in dynamic environments like road traffic, the challenges posed by harsh conditions, nonlinearity, and stochasticity are formidable. This paper introduces a novel approach that synergizes Physics-Informed Neural Networks (PINNs) and Ensemble Transfer Learning (ETL) to address these challenges, enhancing the accuracy and reliability of time series analysis and prediction. PINNs, by incorporating domain knowledge through partial differential equations (PDEs), enable the integration of underlying physics principles into neural network architectures. This fusion of data-driven insights with physical constraints provides a robust framework for capturing complex relationships in time series data. ETL complements PINNs by leveraging multiple models trained on related datasets, enhancing generalization across scenarios and improving forecasting accuracy. A case study focusing on road traffic data is expected to demonstrate the effectiveness of this concept, utilizing real-world traffic data and encoding basic traffic flow equations with PINNs. The anticipated results suggest that the ensemble of PINNs with transfer learning will surpass traditional methods, exhibiting superior predictive capabilities and adaptability to dynamic conditions, even in unobserved scenarios.

Keywords: PINN · Ensemble Learning · traffic forecasting · time-series analysis

1 Introduction

Physics-Informed Neural Networks (PINNs) have emerged as a promising way for the analysis and forecasting of traffic and weather data, ushering in a

H. Unger and M. Schaible (Eds.): AUTSYS 2023, LNNS 1009, pp. 110–121, 2024.
https://doi.org/10.1007/978-3-031-61418-7_5

new era of enhanced accuracy and interpretability. These neural networks, uniquely designed to seamlessly integrate physical principles into their architecture, address the inherent challenges of complex and dynamic systems. In the realms of traffic and weather data, where nonlinear interactions and stochastic behaviors abound, PINNs play a pivotal role in transcending the limitations of traditional methods. By combining the power of data-driven learning with the precision of physics-based constraints, PINNs provide a holistic understanding of the underlying dynamics, leading to more accurate predictions and insightful analyses. We will explore the profound impact of PINNs in revolutionizing the way we comprehend and predict traffic patterns and atmospheric conditions, setting a course for advancements that transcend the boundaries of traditional approaches in time-series data analysis. This paper is our starting point for implementing a new framework that can be used in many different applications that deal with harsh, nonlinear and stochastic time series data.

2 Physics-Informed Neural Networks (PINN): Mechanics, Recent Extensions, and Potentials

Physics-Informed Neural Networks (PINN) represent a powerful paradigm that seamlessly integrates the strengths of physics-based models and neural networks. We explore the mechanics of PINN, discuss some recent extensions, and elucidate their great potential to address complex problems across various domains.

Physics-Informed Neural Networks (PINNs) offer several advantages over traditional neural networks. One key advantage lies in their ability to seamlessly incorporate domain-specific knowledge and physical principles into the learning process. By embedding physics-based constraints through specialized loss functions, PINNs enhance model interpretability and ensure that predictions align with fundamental laws governing the underlying phenomena. This unique feature promotes transparency in model outputs and allows for more robust and reliable predictions, especially in scenarios where adherence to physical laws is crucial. Moreover, PINNs excel in situations with limited or noisy data, leveraging both observed data and prior knowledge to generalize more effectively. This adaptability to diverse datasets, coupled with the capacity to capture complex, nonlinear relationships, positions PINNs as powerful tools for scientific and engineering applications, where traditional neural networks may need help to incorporate domain expertise and ensure adherence to governing principles. The mechanics of Physics-Informed Neural Networks and recent extensions open up exciting possibilities [4,5] for solving complex problems in science and engineering. Fusing neural networks with physics-based principles provides a robust, interpretable, and efficient approach to tackle challenges that traditional methods struggle to address.

2.1 Basics of Physics-Informed Neural Networks

Neural networks' capacity to learn complex patterns and relationships from data has proven invaluable in various scientific disciplines. In the context of physics, however, traditional neural networks lack the ability to incorporate fundamental physical laws. PINN [14] emerges as a solution to bridge this gap as is shown in Fig. 1.

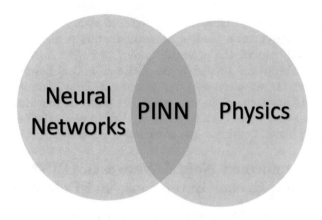

Fig. 1. Physics-informed Neural Networks (PINNs) rely between the traditional neural networks and the physics domain

PINN incorporates physics into the learning process by embedding governing equations directly into the neural network architecture. This is achieved by including physics-based loss terms and enforcing the model to adhere to fundamental principles while learning from data. The training of PINN as shown in Fig. 2 involves minimizing a combined loss function comprising both data-driven and physics-driven terms. This dual-loss structure enables the model to fit the observed data simultaneously and adhere to the underlying physical laws, facilitating robust and accurate predictions.

2.2   Recent Extensions of PINN

Recent advancements have extended PINN to handle multi-physics problems where multiple physical phenomena coexist. These extensions incorporate coupled equations, allowing PINN to simulate and predict complex interactions between physical processes [1,8]. Uncertainty quantification is crucial in many applications, especially in scenarios with limited data or model uncertainties. Recent developments in PINN include methods for estimating and incorporating uncertainties, making predictions more reliable, and providing a measure of confidence in the model outputs. PINN has demonstrated exceptional capabilities in solving inverse problems, where the goal is to infer model parameters or initial/boundary conditions from observed data [2]. Recent extensions involve

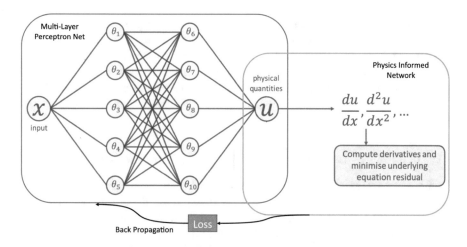

Fig. 2. The training process of a PINN

enhancing PINN's performance in these scenarios, providing a versatile tool for parameter estimation and system identification.

2.3   Potentials of PINN in Solving Complex Problems

PINN excels in handling high-dimensional and nonlinear systems, showcasing their potential in scenarios where traditional numerical methods struggle. The ability to capture intricate relationships between variables positions PINN as a promising tool in fields such as fluid dynamics, heat transfer, and structural mechanics. PINN's incorporation of physics enables it to generalize well, even with limited data. This feature is particularly advantageous in situations where data collection is expensive, time-consuming, or challenging [10]. PINN's ability to leverage physics-based constraints reduces the need for extensive datasets. Another feature that allows it to be applied across diverse scientific domains is PINN's versatility. Whether in aerospace engineering, medical imaging, or material science, PINN offers a unified framework capable of addressing many problems, making it an attractive choice for interdisciplinary research.

3   Challenges in Analyzing and Forecasting Complex and Multivariate Harsh Time Series: Focusing on Traffic and Weather

Effectively analyzing and forecasting complex and multivariate harsh time-series data, particularly in domains like traffic and weather, presents a formidable challenge. Embracing the challenges posed by strong nonlinearity and stochasticity requires adopting advanced methodologies to navigate the intricate relationships within these dynamic systems. We will explore a brief, comprehensive overview of the difficulties associated with traditionally approaching such harsh time-series datasets, emphasizing the intricate interplay between strong non-linearity and stochasticity. As the field continues to evolve, the integration of machine learning, nonlinear models, and a nuanced understanding of the interplay between different variables offers a pathway to overcoming the limitations of traditional methods and unlocking more accurate predictions in complex time-series domains.

3.1   Challenges in Traditional Analysis of Time Series

Traffic systems exhibit nonlinear behavior due to the complex interactions between various factors, such as vehicle density, road conditions, and driver behavior. Traditional linear models often fail to capture these intricate nonlinearities, leading to inaccurate predictions and analyses. Weather patterns are inherently stochastic, with multiple variables influencing each other in dynamic ways. Traditional methods need help to account for the randomness and unpredictability inherent in weather time-series data, hindering their ability to provide accurate forecasts. Both traffic and weather data are multivariate, involving numerous interdependent variables. Traditional methods that assume independence or overlook intricate dependencies between variables fail to capture the holistic nature of these systems, leading to incomplete analyses.

3.2   Challenges in Forecasting Time Series

Traffic forecasting requires accounting for nonlinear dynamics arising from factors like traffic signals, congestion patterns, and unexpected events. Traditional forecasting methods based on linear regression or autoregressive models struggle to capture the complexities, resulting in suboptimal predictions. Weather conditions exhibit temporal and spatial variabilities that traditional models may not adequately address. The nonlinear relationships between atmospheric variables and their evolving patterns challenge the capability of linear forecasting methods to provide accurate and timely predictions.

3.3   Special Considerations for Traffic and Weather Time Series

Weather conditions profoundly impact traffic, influencing variables such as road surface conditions, visibility, and driver behavior. Traditional approaches often need to pay more attention to these interdependencies, leading to inadequate models that fail to capture the nuanced relationship between weather and traffic. Both traffic and weather data are characterized by their sheer volume and complexity. Traditional methods, constrained by data quality and quantity assumptions, may falter when faced with incomplete, noisy, or sparse datasets, limiting their effectiveness in capturing the true nature of the phenomena.

3.4   The Need for Advanced Approaches

Advanced approaches such as machine learning models, neural networks, and other nonlinear techniques have gained prominence to address the challenges posed by strong nonlinearity and stochasticity in traffic and weather time-series data. These methods can capture complex patterns and dependencies, offering more accurate analyses and forecasts. Incorporating physics-based principles into data-driven models represents a promising avenue. Models that blend the strengths of physics-informed techniques with data-driven flexibility can better capture the intricate dynamics of traffic and weather systems, providing more robust and interpretable results.

4   Technical Review of Advanced Physics-Informed Neural Networks for Reliable Analysis and Forecasting of Complex Harsh Time Series in Traffic and Weather

Advanced Physics-Informed Neural Networks have a great application potential for tackling the challenges encountered in ensuring reliable analysis and forecasting of complex, harsh time-series data, specifically in the domains of traffic and weather. We make a review with focus on classification, anomaly detection, and forecasting tasks. Integrating physics with neural networks enhances the robustness of models, providing a more reliable framework for tasks such as classification, anomaly detection, and forecasting. The reported findings from related works consistently demonstrate the superior performance of advanced PINN compared to traditional approaches, highlighting the promise of these models in advancing the state-of-the-art in time-series data analysis and forecasting applications.

4.1   Advanced Physics-Informed Neural Networks

Advanced PINN leverages the integration of physical principles into neural network architectures. This unique combination enables the model to learn from data while adhering to fundamental physical laws, providing a robust framework for analyzing and forecasting complex time-series data affected by nonlinearity and stochasticity [15]. PINN's ability to blend data-driven learning with physics-based constraints offers a unique advantage for ensuring the reliability of predictions [9]. Hybrid models that harness the power of both neural networks and physics-driven constraints are particularly promising for addressing the challenges posed by complex time-series data.

4.2   Classification and Anomaly Detection in Traffic Data

Recent works have explored the application of advanced PINN for traffic flow classification. By incorporating the physics of traffic dynamics into the neural network, these models demonstrate improved accuracy in classifying different traffic patterns, enabling more reliable traffic analysis. Anomalies in traffic data, such as accidents or unexpected congestion, pose challenges for traditional methods. Advanced PINN, with their ability to capture complex relationships and constraints, show promise in anomaly detection tasks, providing more reliable alerts for abnormal traffic conditions.

4.3   Forecasting in Traffic and Weather

Advanced PINN exhibits superior capabilities in forecasting traffic flow by considering the nonlinear dynamics inherent in traffic systems. These models outperform traditional methods, offering more reliable predictions for short-term and long-term traffic patterns. Weather conditions significantly impact traffic, making accurate forecasting challenging. Advanced PINN, by integrating weather data with traffic dynamics, achieves more reliable predictions, capturing the intricate relationships between weather variables and traffic patterns.

4.4   Performance Comparison with Traditional Approaches

Studies comparing advanced PINN with traditional methods highlight the superior accuracy achieved by PINN in traffic analysis tasks. The ability to capture nonlinear dynamics and incorporate physics-based constraints results in more reliable insights into traffic behavior. In forecasting scenarios, advanced PINN consistently outperforms traditional models, especially in the presence of strong nonlinearity and stochasticity. The reliability of predictions is notably improved, offering more accurate insights into future traffic and weather conditions.

5   Ensemble Learning for Enhanced Performance of Physics-Informed Neural Networks in Complex Time Series Analysis and Forecasting

This chapter explores the synergy between Ensemble Learning techniques and Physics-Informed Neural Networks to enhance the overall performance in analyzing and forecasting harsh, complex time-series data [10]. Ensemble methods, including bagging, stacking, and boosting, offer a promising avenue for improving the robustness and accuracy of PINN models. Additionally, the chapter delves into the potential of combining these ensemble strategies with selected Deep Learning (DL) models such as Extreme Learning Machines (ELM), Multi-Layer Perceptrons (MLP), Cellular Neural Networks (CNN), Graph Attention Networks (GAT), and Long Short-Term Memory (LSTM). The discussion briefly explores meaningful architectures and the motivation behind their selection.

5.1   Ensemble Learning Techniques and PINN

There are three famous ensemble learning techniques that are useful when combined with PINNs: Bagging [11], Stacking [12] and Boosting [13].

- Ensemble **bagging** techniques involve training multiple instances of the same model with different subsets of the training data. In the context of PINN, bagging can enhance stability and robustness by mitigating the impact of outliers and noisy data, leading to more reliable predictions.
- **Stacking** combines the predictions of multiple models, creating a meta-model that leverages the strengths of its constituent models. This approach is beneficial for PINN, enabling the fusion of different perspectives and enhancing the overall accuracy of the ensemble in capturing complex time-series patterns.
- **Boosting** focuses on sequentially improving the weaknesses of individual models and adapting them to perform better on specific aspects of the data. For PINN, boosting can enhance adaptability to nonlinearities and stochasticities, resulting in more accurate analysis and forecasting.

5.2   Ensemble Architectures with Selected DL Models

1. **ELM-PINN Ensemble:** Combining PINN with Extreme Learning Machines (ELM) in an ensemble can capitalize on ELM's fast learning capabilities and PINN's ability to incorporate physics-based constraints. This architecture is well-suited for cases where rapid learning and adherence to physical laws are crucial.
2. **MLP-Stacked PINN:** A Stacked Ensemble of PINN with Multi-Layer Perceptrons (MLP) provides a robust approach to capturing complex relationships. The MLP's ability to model intricate patterns complements PINN's physics-driven structure, resulting in a powerful ensemble for time-series analysis and forecasting.
3. **Cellular Neural Network (CNN)-Boosted PINN:** Boosting PINN with Cellular Neural Networks (CNN) enhances adaptability to spatial and temporal dependencies in time-series data. This architecture is particularly beneficial for applications where the interplay between neighboring data points, such as traffic flow dynamics, is crucial.
4. **GAT-Bagged PINN:** Ensembling PINN with Graph Attention Networks (GAT) using a bagging approach is effective when dealing with time-series data exhibiting graph-like structures. GAT's attention mechanisms combined with PINN's physics constraints provide a powerful ensemble for capturing complex interactions.
5. **LSTM-Boosted PINN:** Boosting PINN with Long Short-Term Memory (LSTM) networks is advantageous for scenarios where capturing long-term dependencies is critical. The LSTM's memory cells enhance PINN's ability to model temporal patterns, making it well-suited for forecasting tasks in time-series data.

Each selected architecture combines the strengths of PINN with a specific DL model to address different facets of complex time-series data. ELM provides rapid learning, MLP captures intricate patterns, CNN handles spatial dependencies, GAT deals with graph-like structures, and LSTM captures long-term dependencies. The ensemble architectures aim to achieve robustness by combining models that excel in different aspects of data analysis. Additionally, boosting and stacking strategies enhance adaptability to various challenges posed by harsh, complex time-series data.

6 Enhancing Explainability and Interpretability in Time Series Analysis and Forecasting by Physics-Informed Neural Networks

The integration of Physics-Informed Neural Networks (PINN) in Deep Learning (DL) based Time-Series (TS) analysis and forecasting introduces a unique synergy that enhances both explainability and interpretability. We delve into how PINN contributes to making the often complex and opaque nature of DL models more understandable, transparent, and accessible. By explicitly incorporating physical laws and constraints into the learning process, PINN contributes to making complex models more transparent and interpretable. As research in this area progresses, the continued exploration of model interpretability will be essential for ensuring the widespread adoption of advanced DL techniques in real-world applications.

6.1 Explainability in Time Series Analysis

Traditional DL models, often viewed as black boxes, lack transparency in their decision-making processes. PINN's incorporation of physics-based principles introduces a level of transparency, making it easier to understand how the model arrives at its predictions. This attribute is crucial for establishing trust in the model's outputs. PINN relies on loss functions that incorporate physical laws or constraints. This explicit inclusion of domain knowledge enables stakeholders to understand better how the model aligns with the underlying principles governing the time-series data. A transparent and interpretable loss function enhances the explainability of the model.

6.2 Interpretability in Time Series Forecastings

PINN incorporates physics-driven features, effectively guiding the neural network towards relevant and interpretable features. PINN implicitly performs feature engineering by aligning with physical laws, making the model's decision-making process more understandable. The predictions generated by PINN are not solely based on learned patterns from data but are also constrained by the physical laws embedded in the model architecture. This dual nature of prediction contributes to transparent and interpretable model outputs, allowing users to trace predictions back to fundamental principles.

6.3   Case Studies: Examples of Explainable and Interpretable Outcomes

We consider two case studies of harsh time-series data.

- Traffic Flow Dynamics: In traffic analysis, PINN can explicitly consider traffic laws and dynamics in the loss function, leading to models that not only accurately predict traffic patterns but also provide insights into the factors influencing these predictions. This transparency is vital for urban planning and traffic management.
- Weather-Induced Time-Series: PINN can incorporate meteorological equations in weather-related time-series forecasting, providing interpretable insights into the relationship between weather variables and their impact on the time-series data. This explicit consideration of physical laws enhances the interpretability of weather forecasts.

6.4   Bridging the Gap Between Data-Driven and Physics-Informed Models

PINN seamlessly integrates data-driven learning with physics-based constraints. This dual learning paradigm not only enhances model performance but also ensures that the model's outputs adhere to known physical laws. This combination contributes to both explainability and interpretability. PINN also, allows for flexibility in modeling while maintaining transparency. Users can introduce physics-based constraints specific to the problem domain, providing a level of customization that is often lacking in purely data-driven approaches.

6.5   Challenges and Future Directions

The challenge of PINN lies in balancing the complexity of the model with the need for simplicity in explanation. Striking the right balance ensures that the model remains sufficiently complex to capture intricate patterns while maintaining an understandable and interpretable nature. While PINN contributes significantly to explainability within its framework, future research could explore methods to make the insights derived from PINN more model-agnostic, facilitating broader interpretability across different domains and models.

7   Advantages of Using PINN for Traffic and Weather Data

Many papers like [1–9] leverage PINNs to integrate physical principles into neural network architectures for enhanced predictive capabilities. The application domains include urban flow prediction, traffic flow estimation, traffic density reconstruction and many more. The results across the papers consistently indicate the potential of PINN to outperform traditional methods and provide accurate, physics-informed predictions in their respective domains. PINN's unique strength lies in its ability to seamlessly integrate physics-driven constraints with data-driven learning, making it suitable for applications where capturing complex interactions and adherence to physical laws are crucial. The advantages of PINNs across these papers include improved accuracy, physics-informed predictions, handling uncertainty, accurate simulation, and the ability to model complex physical interactions in harsh time-series data.

8 Future Work

In the utilization of real-world traffic data, Physics-Informed Neural Networks (PINNs) prove to be instrumental by encoding fundamental traffic flow equations, facilitating the comprehensive analysis of intricate relationships among various traffic variables. The Ensemble Transfer Learning (ETL) framework further enhances predictive capabilities by aggregating insights from multiple PINN instances. Each of these instances is trained on distinct subsets of data, effectively capturing diverse traffic conditions or geographic areas. The incorporation of stochastic knowledge equips the model with superior adaptability, enabling it to effectively capture abrupt changes, traffic congestion, and irregular patterns. This added stochastic knowledge not only improves the model's predictive accuracy but also enhances its adaptability to changing conditions, ensuring that it can provide precise forecasts even in scenarios that were not observed during training. The result is expected to be a robust and versatile framework for different applications of traffic data analysis and forecasting that transcends the limitations of traditional approaches.

References

1. Zhang, J., Mao, S., Yang, L., Ma, W., Li, S., Gao, Z.: Physics-informed deep learning for traffic state estimation based on the traffic flow model and computational graph method. Inf. Fusion **101** (2024)
2. Wong, J., Chiu, P., Ooi, C., Da, M.: Robustness of physics-informed neural networks to noise in sensor data (2022)
3. Barreau, M., Aguiar, M., Liu, J., Johansson, K.H.: Physics-informed learning for identification and state reconstruction of traffic density. In: 2021 60th IEEE Conference On Decision And Control (CDC), pp. 2653–2658 (2021)
4. Mo, Z., Fu, Y., Xu, D., Di, X.: TrafficFlowGAN: physics-informed flow based generative adversarial network for uncertainty quantification. In: Amini, M.R., Canu, S., Fischer, A., Guns, T., Kralj Novak, P., Tsoumakas, G. (eds.) ECML PKDD 2022. LNCS, vol. 13715, pp. 323–339. Springer, Cham (2023). https://doi.org/10. 1007/978-3-031-26409-2_20
5. Koeppe, A., Bamer, F., Selzer, M., Nestler, B. Markert, B:. Explainable artificial intelligence for mechanics: physics-explaining neural networks for constitutive models. Comput. Mater. Sci. (2021)
6. Raissi, M., Perdikaris, P., Karniadakis, G.: Physics-informed neural networks for urban flow prediction (2017)
7. Riccius, L., Agrawal, A. Koutsourelakis, P.: Physics-informed tensor basis neural network for turbulence closure modeling (2023)

8. Sel, K., Mohammadi, A., Pettigrew, R.I., et al.: Physics- informed neural networks for modeling physiological time series for cuffless blood pressure estimation. NPJ Digit. Med. **6** (2023)

9. Tang, H., Liao, Y., Yang, H., et al.: A transfer learning-physics informed neural network (TL-PINN) for vortex-induced vibration. Ocean Eng. **266** (2022)

10. Chakraborty, S.: Transfer learning based multi-fidelity physics informed deep neural networks. J. Comput. Phys. **426** (2021)

11. Breiman, L.: Bagging predictors. Mach. Learn. **24**, 123–140 (1996)

12. Wolpert, H.: Stacked generalization. Neural Netw. **5**, 241–259 (1992)

13. Freund, Y., Schapire, E.: A decision-theoretic generalization of on-line learning and an application to boosting. J. Comput. Syst. Sci. **55**, 119–139 (1997)

14. Cuomo, S., Di Cola, V.S., Giampaolo, F., Rozza, G., Raissi, M., Piccialli, F.: Scientific machine learning through physics-informed neural networks: where we are and what's next. J. Comput. Syst. Sci. (2022)

15. Raissi, M., Perdikaris, P., Karniadakis, E.: Physics-informed neural networks: a deep learning framework for solving forward and inverse problems involving nonlinear partial differential equations. J. Comput. Phys. **378**, 686–707 (2019)

Language Meets Vision: A Critical Survey on Cutting-Edge Prompt-Based Image Generation Models

Paraskevi Fasouli[✉], Witesyavwirwa Vianney Kambale,
and Kyandoghere Kyamakya

Institute for Smart Systems Technologies, Universität Klagenfurt,
Universitätsstraße 65/67, 9020 Klagenfurt, Austria
Paraskevi.Fasouli@aau.at

Abstract. In the dynamic landscape of computer vision and artificial intelligence, the fusion of language and image generation has emerged, giving rise to Language-Driven Image Generation Models. This comprehensive paper navigates the intricate realm of language-driven image generation models. Beginning with a comprehensive specification book, this paper offers guidelines for practitioners and researchers in the domain of prompt-based generative models. A historical overview reviews the evolution of deep learning in image generation, providing valuable context for the subsequent comparative analysis of State-of-the-art models. This analysis critically evaluates leading models, identifying gaps and fostering a nuanced understanding of their capabilities. The exploration concludes with a focus on training techniques for generative models, shedding light on challenges and potential avenues for refinement. In essence, this paper serves as a comprehensive guide, steering readers through the evolution, intricacies, and future directions of Language-Driven Image Generation Models, fostering a deeper understanding and encouraging continued exploration in this dynamic interdisciplinary field.

Keywords: Prompt-based image generation · Generative models · Large Language Models · Fine-tuning

1 Introduction

In recent years, the intersection of language and image generation has witnessed remarkable advancements, giving rise to a revolutionary domain in computer vision known as Language-Driven Image Generation Models. This paper embarks on a multifaceted exploration, beginning in chapter two with a comprehensive description of the concept. As we delve into the intricacies of prompt-based generative models, the third chapter unfolds with the presentation of a comprehensive specification book designed to serve as a guiding compass for practitioners and researchers. Building on this foundation, the paper conducts a thorough historical overview in the forth chapter, reviewing the evolution of deep learning

© The Author(s), under exclusive license to Springer Nature Switzerland AG 2024
H. Unger and M. Schaible (Eds.): AUTSYS 2023, LNNS 1009, pp. 122–140, 2024.
https://doi.org/10.1007/978-3-031-61418-7_6

in image generation, providing a contextual backdrop for the subsequent critical comparative analysis of the State-of-the-art models in the fifth chapter. This analysis scrutinizes leading models, offering a nuanced evaluation and paving the way for a deep dive into training techniques for generative models at the sixth chapter. This holistic exploration aims to unravel the complexities, potentials, and challenges inherent in the fusion of language and image generation, contributing to the collective understanding and advancement of this captivating field.

2   Description of the Concept of Language-Driven Image Generation Models

Prompt-based image generation represents a fundamental transformation of how artificial intelligence creates visual content, seamlessly marrying the expressive power of natural language with the visual creativity of generative models. This concept embodies using textual prompts or descriptions as guiding instructions for generating realistic and often highly detailed images. Unlike traditional approaches, where input was primarily fixed data sets or predefined labels, prompt-based image generation opens a dynamic channel for users to articulate their creative visions. Language models are important to generative models as they determine how semantically related a natural language snippet is to a visual concept, which is critical for text-conditional image generation. The text prompts act as the image descriptors of the final image. The model interprets and translates these textual cues into visually compelling outputs, demonstrating an intersection of language understanding and image synthesis.

2.1   Background and Evolution of Prompt-Based Image Generation

Early attempts to integrate language and image generation date back several decades, with researchers exploring ways to bridge the gap between natural language understanding and computer vision. While these efforts were often rudimentary compared to contemporary advancements, they laid the groundwork for developing more sophisticated language-driven image generation models. Some notable projects that contributed to the foundational concepts of natural language understanding that later influenced integrated AI systems are:

- **Conceptual Dependency Theory (CDT):** Introduced in the 1970s by Roger Schank [29], CDT focused on representing knowledge in a form that machines could understand. This concept influenced early work on knowledge representation and language understanding.
- **SHRDLU:** Developed by Terry Winograd at the Massachusetts Institute of Technology (MIT), SHRDLU [28] was an early natural language understanding system. It operated in a block-world environment, where a set of blocks could be manipulated based on textual commands. The system generated simple visual representations in the block world, allowing users to describe actions and queries using natural language. The system would respond by manipulating the blocks accordingly.

- **CYC Project:** Initiated by Douglas Lenat [27] in 1984, this project aimed to create a comprehensive knowledge base that could be used for reasoning and understanding natural language. It was a core contribution to the foundational concepts that later influenced integrated AI systems.
- **Generative Interactive Virtual Environment (GIVE):** Initiated in the 1990s, GIVE [26] was a project that focused on creating a virtual environment where users could interact using natural language. While primarily aimed at language understanding and interaction, it laid the groundwork for future research into combining language and visual elements in virtual environments.

The limitations of computing power, available data sets, and the understanding of natural language processing constrained these early attempts. Text-to-image generation gained significant popularity again and experienced a surge in research interest in the mid-2010s, particularly with the advent of deep learning techniques.

2.2 Motivation and Expectations of Prompt-Based Image Generation

The motivation for employing language-driven approaches in image generation stems from the natural collaboration between natural language and visual representation. Researchers and practitioners desire to explore the fusion of language and image synthesis because it creates better collaboration between human and AI systems. Language-driven image generation aims to align with human cognitive processes, facilitating a more intuitive and user-friendly interaction. Leveraging language as a driving force, enables a closer cognitive alignment between user intent and model output, enhancing the overall user experience. The motivation to simplify the content creation process for individuals without specialized design skills has driven the exploration of language-driven models. AI systems can become more user-friendly and offer a more expressive communication with their users. Another motivation is the contextual understanding and enhanced creativity that it offers. Language-driven models aim to capture and leverage contextual information embedded in linguistic prompts. This is essential for generating visually coherent and contextually relevant images that align with user expectations. Users can then steer their creative process and tailor the visual output to their specific requirements by providing more detailed prompts.

Using language as a driving force in image creation, users can expect several advantages over traditional methods, introducing new possibilities and enhancing the overall process of image synthesis. One key advantage is its user-friendly and intuitive interaction, allowing individuals to articulate their creative visions effortlessly through natural language prompts. This results in expressive and specific image synthesis, where users can precisely describe their desired visual output. The context-aware image synthesis capability ensures that the generated images are coherent and contextually relevant, enhancing the overall user experience. Moreover, the system's adaptability to various styles and genres allows users to explore and create content across various artistic expressions.

Fine-grained control and specificity in the generation process provide users with creative granularity, fostering a personalized approach to image creation. The model's ability for domain-specific image generation and creation ensures relevance and accuracy in specialized fields. Furthermore, incorporating multimodal input processing enables users to combine textual and visual prompts for a richer and more versatile interaction. This facilitates creative exploration and makes content creation accessible to a broader audience. In summary, prompt-based image generation stands out for its versatility, user-centric design, and the ability to cater to diverse creative needs.

3   Comprehensive Specification Book for Prompt-Based Generative Models

A detailed specification book is invaluable in the dynamic landscape of language-driven models, delineating the foundational principles and intricate details about their design, functionality, and user interactions. This involves a thorough examination of the input specifications, the role of context, and the intricacies of interpreting linguistic prompts. We discuss structuring and designing prompt-based tasks to ensure meaningful and coherent visual outputs. Furthermore, we explore the technical aspects of implementing language-driven models, such as the architecture considerations, training methodologies, and model interpretability.

3.1   User Input (Natural Language Prompts) Guidelines

Natural language prompts form the core of communication between users and the language-driven image generation model. Prompts are natural language text describing an AI's task, which can be a query, a command, a short statement of feedback, or a longer statement including context, instructions, and input data [20]. Prompt engineering involves carefully designing and crafting the prompts or instructions given to the model to elicit desired responses [17]. The presented guidelines aim to assist users in formulating effective natural language prompts for the language-driven image generation model. Users can enhance the model's understanding by being descriptive, avoiding ambiguity, providing contextual cues, and generating more accurate and desirable visual outputs. The guidelines can be classified into two categories: (i) Specificity and Detail, (ii) Contextual Cues.

Specificity and Detail. Encouraging users to provide specific and detailed prompts is crucial to guide the model accurately for achieving desired results. Three good ways of generating text prompts that provide accurate results are by (i) being descriptive and clear, by (ii) avoiding ambiguity and by (iii) specifying arrangement and composition. Some examples can be:

Example (i) Unclear, generic prompt: *"Mountain landscape"*, Descriptive and clear prompt: *"A panoramic view of snow-capped mountains with a clear blue sky and a winding river"*.

Example (ii) Ambiguous prompt: *"Abstract art"*, Better specified prompt: *"Create an abstract and dynamic composition with vibrant geometric shapes and contrasting colors"*.

Example (iii) Generic prompt: *"City skyline"*, Prompt with specified arrangement, *"A nighttime city skyline similar to New York, with tall skyscrapers illuminated by city lights"*.

Figure 1 shows the results of the image generation of the ***example (i)*** prompts. The model used was Stable Diffusion version (XL 1.0) [3] with CLIP for the prompt-based image generations. The left image is a more abstract and generic mountain landscape, whereas the right image that follows the more descriptive prompt, shows a more guided generated landscape. The mountains have indeed a panoramic view, snow on their top and there is a river passing through them.

"Mountain landscape"   "A panoramic view of snow-capped
                                  mountains with a clear blue sky and a
                                  winding river"

Fig. 1. The resulted images of the prompts for a mountain landscape

Contextual Cues. Contextual cues help the model understand the intended context of the image [16,17,20,21]. Two important things to consider on incorporating contextual information in prompts to enhance the model's contextual awareness are to provide context for better interpretation (i) and to consider spatial relationships (ii) between elements in the image to enhance coherence. Some examples can be:

Example (i) Prompt with unclear interpretation: *"Modern kitchen"*, prompt with added context: *"Design a modern kitchen with sleek counter tops, stainless steel appliances, and large windows"*.

Example (i) Lack of contextual cue prompt: *"Futuristic city"*, contextual cue Prompt: *"Create a futuristic cityscape with skyscrapers and flying cars reminiscent of a sci-fi movie"*.

Example (ii) Prompt without spatial relationships: *"Forest scene"*, better formed prompt: *"A dense forest with sunlight filtering through tall trees, creating dappled shadows on the forest floor"*.

Figure 2 shows the results of the image generation of the ***example (ii)*** prompts. The model used was Stable Diffusion version (XL 1.0) [3] with CLIP for the prompt-based image generations. The left image is a more abstract and generic forest landscape, whereas the right image follows the more detailed description and shows a more guided forest that depicts the described details of the sunlight and the tall trees.

"Forest scene"   "A dense forest with sunlight filtering through tall trees, creating dappled shadows on the forest floor"

Fig. 2. The resulted images of the prompts for a forest landscape

3.2   Training Data

It is essential to specify the sources of the training data, ensuring transparency about the diversity and domains covered. This is due to the fact that training data are one of the most important factor for a model's accuracy as they act as the knowledge base of it. Domain-specific considerations are crucial when the intended application or user requirements demand expertise in a particular field. For instance, if the language-driven model is designed for architectural visualization, the training data should include ample examples of architectural scenes, building styles, and contextual details. Similarly, the model needs exposure to medical imagery and terminology for medical imaging applications.

In selecting domain-specific training data, addressing ethical considerations and potential biases is imperative. Ensuring *diversity* and *fairness* in the data set prevents the model from unintentionally perpetuating stereotypes or exhibiting biased behavior and the generated images can be inclusive and representative of different perspectives. Another important factor when arranging training data is respecting the rights and consent of individuals whose images are included in the data set. *Ethical considerations* demand careful handling of sensitive content to avoid potential harm or discomfort to individuals depicted in the images. Lastly, *transparency* builds trust and allows users and developers to understand the origins of the data, promoting accountability. It is recommended to use data sets that provide transparent documentation of the sources of the training data, including image repositories.

3.3 Model Capabilities

When choosing a model for generating content, it is important to consider the styles and genres of the intended output, the resolution and the quality of it and finally the model architecture. The styles and genres a language-driven image generation model is trained to handle can vary based on the data set and training objectives. According with the use case application, there are various styles to select. Some examples can be: Landscape painting, Abstract art, Sci-fi illustrations, Portrait photography, Minimalist design, Impressionist art, Architectural visualisation, comic book illustrations, Retro poster design, Fantasy landscapes etc. The resolution and quality of the generated images should be carefully considered according to an trade-offs or limitations that can occur due to the hardware or the model architecture or the quality of training data.

A language-driven image generation model typically involves a combination of a natural language processing (NLP) model and an image generative model. A brief description of the key components of such a model is presented.

1. *Text Embedding:* The natural language prompt is initially processed through a text embedding layer. This layer converts the input text into a numerical representation, often using pre-trained word embeddings.
2. *Encoder-Decoder Architecture*: The model generally follows an encoder-decoder architecture [20]. The text embedding serves as the input to an encoder, which processes and encodes the information. The encoder captures semantic features from the input text. The encoded information is then fed into the decoder, which is responsible for generating the corresponding image. The decoder translates the encoded text into visual features, gradually constructing the image through a series of layers.
3. *Attention Mechanism:* Many advanced language-driven image generation models incorporate attention mechanisms. Attention mechanisms [16] allow the model to focus on specific parts of the input text when generating different regions of the image. This enhances the model's ability to align visual details with corresponding textual cues.

4. *Image Generation Layers:* The final layers of the model are responsible for generating pixel-level details in the image. These layers translate the abstract features extracted from the text into concrete visual elements.

5. *Fine-Tuning Mechanisms:* Many language-driven image generation models incorporate fine-tuning mechanisms. These mechanisms adjust the model based on feedback, either from users or through automated evaluation metrics, to improve the quality and relevance of generated images over time.

6. *Conditional Image Synthesis:* The entire model operates conditionally, taking the input text as a condition for generating the image. This conditionality allows users to guide the model's creativity and specify the desired visual content.

7. *Transfer Learning and Pre-training:* Many models leverage transfer learning and pre-training on large-scale data sets [13]. Pre-training on diverse data sets helps the model learn general features, and then fine-tuning on a more specific data set tailors the model to the task at hand.

3.4 Output Evaluation

User satisfaction is the ultimate goal, ensuring that the language-driven image generation model enhances the user's creative experience and effectively translates their ideas into compelling visuals. A well-evaluated image should meet technical standards and resonate with the user's creative vision, preferences, and expectations. For the model to create high-quality, relevant images, it should comply with the following proposed criteria.

1. *Visual Fidelity:* Visual fidelity refers to the degree of accuracy in representing the details specified in the user's prompt. It assesses how well the generated image captures the visual elements described, including colors, shapes, and textures. High visual fidelity ensures that the generated image closely aligns with the user's creative vision, enhancing the overall quality of the output.

2. *Relevance to Prompt:* This criterion evaluates how closely the generated image aligns with the user's original prompt. It considers whether the model accurately interpreted the user's instructions and produced a relevant visual output. Relevance to the prompt ensures the model understands and responds appropriately to user instructions, minimizing the risk of generating irrelevant images [18].

3. *Creativity and Originality:* Assessing the creativity and originality of the generated image involves considering unique elements, artistic expression, and deviations from generic or expected outputs. Creativity ensures that the model goes beyond reproducing common patterns, providing users with novel and imaginative visual interpretations that align with artistic intent.

4. *Composition and Layout:* This criterion evaluates the generated image's arrangement and composition of visual elements. It considers factors such as balance, spatial relationships, and overall aesthetics. Attention to design and layout contributes to the generated image's overall coherence and visual appeal, creating a more aesthetically pleasing result.

5. *Contextual Consistency:* Contextual consistency assesses how well the generated image aligns with the implied context or setting described in the user's prompt. It considers whether the visual elements make sense within the specified context. Ensuring contextual consistency enhances the realism and relevance of the generated image, making it more faithful to the user's intended scenario or theme [20].
6. *Diversity in Output:* This criterion considers the diversity of outputs generated by the model for a given prompt. It assesses whether the model can produce a range of visually distinct results. A diverse set of outputs provides users with options and allows them to choose the image that best fits their preferences or requirements.
7. *Iterative Refinement Process:* This process establishes a user feedback loop to address cases where the model misinterprets prompts, ensuring continuous improvement of the model's performance. In combination with improved and more detailed refined text prompt that was discussed before, the model can have a better understanding on the user's vision.

3.5   Training Pipeline

The training procedure of a prompt-based image generation model can be summarized into the steps depicted in Fig. 3.

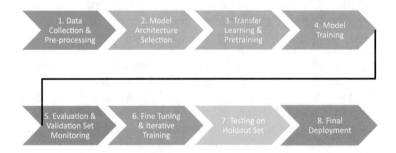

Fig. 3. The 8 steps of the training pipeline of prompt-based image generation

1. *Data Collection and Preprocessing:* Throughout this process, the user should gather a diverse data set that includes paired images and natural language prompts, making sure it represents the styles, genres, and domains that needed for the intended use. Afterwards, the textual data should be tokenized and embedded into numerical representations. Lastly, the images should be resized and normalized to a consistent format.
2. *Model Architecture Selection*: An appropriate architecture should be chosen for the language-driven image generation model. Factors such as the specific task requirements, computational resources, and the availability of pretrained models are of highly consideration. Due to the multimodal nature of

this complex task, 2 models are required, one for the text embedding and one for the image generation.

3. *Transfer Learning and Pre-training:* Leveraging transfer learning by initializing the model with weights from pre-trained models on large-scale data sets is recommended if possible. This step helps the model learn general features before fine-tuning on the specific task.

4. *Model Training*: The model is trained using an encoder-decoder architecture. The encoder processes the textual input, and the decoder generates corresponding visual features. During training, the loss functions that are combined guide the generation towards images that align with the textual prompts.

5. *Evaluation and Validation Set Monitoring:* This step is crucial to assess the model's performance. Typical metrics include visual fidelity, relevance to prompt, and diversity in generated images. The model is monitored with a validation set throughout training. This helps prevent over-fitting and provides insights into when to stop training to avoid performance degradation.

6. *Fine-tuning and Iterative Training*: This step helps adapt the model to specific user preferences and ensures continued improvement over time. Iterative training cycles with periodic updates to the model, incorporate user feedback, adjust hyperparameters, and introduce new training data.

7. *Testing on Holdout Set*: After training is complete, the model should be evaluated on a holdout set or a separate test set to assess its generalization capabilities on unseen data.

8. *Final Deployment*: When considering deployment, it is important to asses the model size, inference speed, and resource requirements. The model should be optimal for deployment while ensuring it meets user expectations.

4   Historical Overview Review of Deep Learning in Image Generation

Text-to-image generation gained significant popularity and experienced a surge in research interest in the mid-2010s, particularly with the advent of deep learning techniques. The inception of deep learning techniques, particularly convolutional neural networks (CNNs), laid the groundwork for image generation tasks. While early attempts at text-to-image synthesis date back to the 1980s, the field saw a renewed focus and acceleration in progress around the following key time-frames:

- *(2014–2015) Deep Learning Resurgence:* The resurgence of interest in deep learning, fueled by advancements in neural network architectures and the availability of large data sets, played a crucial role in the renewed popularity of text-to-image generation. Researchers began leveraging deep generative models, such as Generative Adversarial Networks (GANs) and Variational Autoencoders (VAEs), to tackle the challenges of generating realistic images from textual descriptions.

– *(2015–2016) DCGAN and Visualizing Convolutional Neural Networks:* In 2015, the introduction of Deep Convolutional Generative Adversarial Networks (DCGANs) marked a significant breakthrough in generating high-quality images. DCGANs demonstrated the ability to synthesize visually appealing images, laying the groundwork for more complex text-to-image generation tasks. Around the same time, efforts to visualize and understand the internal representations of Convolutional Neural Networks (CNNs) in computer vision tasks indirectly contributed to advancements in text-to-image synthesis. Researchers realized the potential to manipulate and generate images by influencing the feature spaces (latent space) of CNNs.

– *(2016–2017) StackGAN and Progressive GAN:* In 2016, the StackGAN [24] model was introduced, demonstrating the generation of detailed images in a step-wise manner. This approach involved generating low-resolution images first and progressively refining them to higher resolutions, aligning with the multi-stage nature of textual descriptions. Progressive GANs, introduced in 2017, further improved the stability and image quality of GANs. This contributed to the overall progress in text-to-image generation by enhancing the capabilities of generative models.

– *(2018–2019) BigGAN and Transformers:* In 2018, BigGAN showcased the power of large-scale GANs in generating high-fidelity images. The model, trained on massive data sets, demonstrated the potential for scaling up generative models to achieve remarkable results in image synthesis. The same period saw the rising influence of transformer-based architectures in natural language processing, with models like BERT and Generative Pre-trained Transformer 2 (GPT-2) [7] showcasing the effectiveness of transformers in capturing contextual information. This influenced the integration of transformer architectures in multimodal models for text-to-image synthesis.

– *(2021) CLIP: Connecting Text and Images for Prompt-Based Generation:* OpenAI introduced CLIP (Contrastive Language-Image Pre-training) [6] as a model capable of understanding images and text in a unified framework. CLIP demonstrated the ability to align textual descriptions with corresponding images, opening up new possibilities for prompt-based image generation. This model paved the way for more sophisticated interactions between language and visual content.

– *(2021) DALL-E: Text-to-Image Synthesis with Variational Autoencoders:* OpenAI's DALL-E [2] showcased the power of variational autoencoders in generating diverse and creative images from textual prompts. By learning a latent space that represents diverse visual concepts, DALL-E enabled users to specify novel and complex visual ideas through language, demonstrating the potential for fine-grained control in image synthesis.

– *(2021 - Present) CLIP-Guided Image Generation Models:* Building on CLIP's success, researchers began integrating CLIP into various image generation models, creating CLIP-guided models [12]. These models leverage the semantic understanding of both text and images, allowing users to provide textual prompts that guide the generation of images in a more controlled and interpretable manner.

- *(2022 - Present) Multimodal Transformers & Advancements in Transfer Learning:* The development and exploration of multimodal transformer [11] architectures have played a crucial role in enhancing prompt-based image generation. These architectures integrate transformer models to process both language and visual inputs, allowing for more sophisticated interactions between textual prompts and image synthesis. The application of transfer learning techniques, where models pre-trained on large data sets are fine-tuned for specific tasks, has become a key milestone. This approach allows prompt-based image generation models to leverage knowledge acquired from diverse data sets, resulting in improved performance and adaptability to various domains.
- *(Ongoing) Research in Ethical and Responsible AI:* The recognition of ethical considerations in prompt-based image generation, such as bias in training data and potential misuse, has prompted ongoing research into responsible AI practices. This milestone underscores the importance of addressing societal implications and ensuring the ethical use of generative models in diverse applications.

5 Critical Comparative Analysis of the State-of-the-Art Prompt-Based Image Generation Models

In the landscape of prompt-based image generation, understanding the strengths, weaknesses, and nuances of state-of-the-art models is paramount for advancing the field. This critical comparative analysis of prominent models, aims to provide an insightful examination of their performance, capabilities, and limitations. Through this analysis, we seek to elucidate the current state of prompt-based image generation models, offering a foundation for informed discussions, future research directions, and the continuous refinement of these cutting-edge technologies.

5.1 Review of the State-of-the-Art Models

Prompt-based image generation involves the synthesis of visual content based on textual descriptions. To achieve successful text-to-image generation, a suitable generative model architecture must be chosen. The architecture should align with the specific requirements of the use case. Some of the state-of-the-art generative models for synthesizing an image are the following.

- *Generative Adversarial Networks (GANs):* GANs [24] are characterized by a generative model and a discriminative model working in tandem. The generator creates data instances, and the discriminator evaluates them. This adversarial process enhances the realism of generated data. They excel in generating high-fidelity, realistic images. They are versatile and capable of capturing intricate patterns in data, making them popular for tasks like image synthesis. GANs can be challenging to train and may suffer from mode collapse, where the generator focuses on a limited range of outputs. Ensuring stability during training is an ongoing concern.

– *Variational Autoencoders (VAEs):* VAEs are probabilistic models that combine an encoder and a decoder. The encoder maps input data to a probabilistic latent space, and the decoder reconstructs the input from this latent representation. VAEs offer probabilistic interpretations, allowing for uncertainty estimation in generated samples. They are effective in capturing latent structures and enabling smooth interpolation between data points. VAEs may produce less visually realistic outputs compared to GANs. The deterministic nature of the latent space can limit the diversity of generated samples.

– *Diffusion Models:* Diffusion models [9] frame image generation as the process of iteratively applying noise to an initial image, gradually diffusing it into the desired output. They leverage the mathematical concept of diffusion for generative purposes. Diffusion models excel in capturing complex dependencies in data and are adept at handling uncertainty. They have shown promising results in generating high-resolution images. Training diffusion models can be computationally expensive. The iterative nature of diffusion processes may introduce challenges in terms of inference speed.

Prompt-based image synthesis involves combining natural language processing (NLP) models with image generation models. It is crucial to embed textual descriptions into a format that can be understood by the generative model. Some of the best candidates for prompt generation and understanding are the following language models:

– *GPT (Generative Pre-trained Transformer):* GPT [7] models, or similar transformer-based language models, can be employed to generate coherent and contextually relevant textual prompts. They are versatile and capture contextual information effectively, making them suitable for creative language tasks, including text-based prompts for image synthesis. GPT primarily focuses on language tasks and lacks inherent integration with visual information, limiting its ability to handle tasks that require joint understanding of both text and images.

– *CLIP (Contrastive Language-Image Pre-training):* CLIP [6] is designed for learning joint representations of images and text. It can be used to embed both textual and visual information into a shared space, facilitating effective communication between the language and image components. CLIP's pre-trained representations can be fine-tuned for various downstream tasks, making it adaptable to different applications. While proficient in understanding images and text, CLIP does not directly generate images but can guide other generative models in the process.

– *T5 (Text-to-Text Transfer Transformer):* T5 [5] can be employed for various NLP tasks, including text generation. It allows for flexible text-to-text transformations, making it adaptable to generating textual prompts for image synthesis. T5 can be fine-tuned for different applications, offering a flexible approach to handling diverse language tasks. Like GPT, T5 is primarily focused on language tasks and may lack direct integration with visual information.

– *DistilBERT:* A distilled version of BERT [19], offers a lighter and more efficient alternative. It can be used for encoding textual prompts, reducing computational demands while maintaining performance due to transfer learning inheritance from BERT. Compared to larger models, DistilBERT may struggle with capturing long-range dependencies and intricate context in text. It primarily focuses on language tasks and lacks inherent integration with visual information but it is a good candidate for multimodal integration with an image generation model for generating visuals.

5.2   Comparative Evaluation

The choice between GANs, VAEs, and Diffusion models depends on specific use cases, considering factors such as training stability, latent space representation, image fidelity, and computational complexity. Each model has its strengths and trade-offs, making them suitable for different generative tasks.

- *Training Stability:* GANs often face challenges in training stability, especially avoiding mode collapse. VAEs provide a more stable training process but may sacrifice some visual fidelity. Diffusion models strike a balance, with stability and the potential for high-quality outputs.
- *Latent Space Representation:* VAEs explicitly model a probabilistic latent space, offering insights into uncertainty. GANs have an implicit latent space, while diffusion models work with a deterministic approach, potentially limiting diversity.
- *Image Fidelity:* GANs are renowned for generating visually impressive and sharp images. VAEs may produce less visually striking outputs, and diffusion models aim for a balance between fidelity and diversity.
- *Computational Complexity*: GANs can be computationally intensive, especially in high-resolution image generation. VAEs generally have lower computational demands, and diffusion models, while promising, may also require significant computational resources.

Tables 1, 2 and 3 show the key characteristics, strengths and weaknesses of the 3 model categories discussed above.

Other important criteria that influence the final choice for making a multimodal model can be the inference time, the training time, the image quality and the number of supported styles for the final generation. Table 4 shows some of the most famous state-of-the-art models and compares their performance.

Table 1. Key Characteristics of Image Generation Models

Model Category	Key Characteristics
Generative Adversarial Networks (GANs)	Two models (generator & discriminator) in adversarial training
Varational Autoencoders (VAEs)	Probabilistic encoder-decoder model with latent space
Diffusion Models	Iteratively apply noise to an initial image using diffusion

Table 2. Strengths of Image Generation Models

Model Category	Strengths
Generative Adversarial Networks (GANs)	High-fidelity, realistic outputs; versatile; intricate pattern capturing
Variational Autoencoders (VAEs)	Probabilistic interpretations; effective latent structure; uncertainty estimation
Diffusion Models	Complex dependency capture; potential for high-resolution images

5.3 Gap Analysis: Identifying Limitations

Prompt-based image generation models have demonstrated significant progress, but they still face several notable challenges and limitations. Ambiguity in textual prompts remains a challenge, with models often struggling to interpret vague or unclear instructions, resulting in varied and sometimes unpredictable image outputs. Achieving fine-grained control over specific image features, such as color or shape, is an ongoing challenge, as is handling long-form text descriptions. The diversity and creativity of generated outputs require improvement to move beyond repetitive or stereotypical images. Contextual understanding and memory retention pose difficulties, impacting the model's ability to maintain coherence over extended interactions. Addressing biases inherited from training data and ensuring ethical considerations in generated content is crucial. Evaluating the quality of generated images remains challenging, and metrics may not fully capture perceptual quality or user expectations. Computational resource intensiveness, multimodal integration, real-time generation, and handling rare concepts further underscore the current limitations of prompt-based image generation models. Research efforts are actively addressing these limitations, and future developments are expected to lead to more advanced and capable prompt-based image generation models.

6 Training Techniques for Generative Models

Training techniques play a crucial role in content generation for several reasons, influencing the performance, efficiency, generalization and adaptability of

Table 3. Weaknesses of Image Generation Models

Model Category	Weaknesses
Generative Adversarial Networks (GANs)	Training instability; mode collapse; computation intensity
Variational Autoencoders (VAEs)	Less visually realistic; deterministic latent space
Diffusion Models	Computational complexity; training computational expense; iterative nature

Table 4. Possible combinations of Language and Image generation models

Models	Language Model	Inference Time	Training Time	Styles
VQ-GAN [1]	No	Moderate	Moderate	Data influenced
DALL-E [2]	CLIP	High	High	Prompt influenced
DALL-E 2	GPT [7]	High	High	Prompt influenced
GLIDE [4]	GPT	High	High	Prompt influenced
Stable Diffusion [3]	CLIP [6]	Moderate	High	Data influenced
Imagen [25]	T5 [5]	High	Moderate	Data influenced

the models. They are fundamental for developing models that can effectively converge and transform textual prompts into diverse and contextually relevant visual content. Three key techniques used in the multimodal pipeline for training both image and text generative models are transfer learning, zero-shot learning and fine-tuning.

1. *Transfer learning*: It involves pre-training a model on a large dataset for a related task before fine-tuning it for the specific target task. In prompt-based image generation, transfer learning is often applied to language models like GPT [7], T5 [5], or DistilBERT. These models, initially designed for general language understanding, can leverage their pre-trained knowledge when applied to the specific task of generating images from text. Transfer learning is essential because it allows prompt-based models to benefit from the broad knowledge gained during pre-training on large corpora. The model can then leverage this knowledge to better understand the nuances of textual prompts and improve its image generation capabilities.

2. *Zero-Shot Learning*: Zero-shot learning involves training a model to recognize or generate classes or concepts it has never seen during training. In prompt-based image generation, zero-shot learning allows a model to generate images for concepts that were not explicitly part of the training data. For example, generating images for novel or rare concepts based on textual prompts. Zero-shot learning is crucial because it broadens the applicability of prompt-based models. Instead of being limited to generating images for concepts seen during training, the model can extrapolate its knowledge to generate novel and diverse visual content.

3. *Fine-Tuning*: It is the process of adjusting a pre-trained model on a smaller data set specific to the target task. In prompt-based image generation, fine-tuning is applied to adapt the pre-trained language model to the nuances of generating images from text. The model's parameters are adjusted based on the task-specific data set. Fine-tuning is essential because it tailors the model's knowledge to the specific requirements of prompt-based image generation. It allows the model to specialize in understanding and responding to certain textual prompts, resulting in more accurate and contextually relevant image generation. Example of fine-tuning is the Dreambooth [30] deep learning generation model developped by Google, that is used to fine-tune Stable Diffusion [9] and Imagen [25] models.

6.1   Challenges in Training

Training a model using zero-shot learning, transfer learning, or fine-tuning introduces specific challenges that impact the model's performance and adaptability. In zero-shot learning, where the model is expected to generalize to unseen classes, challenges arise in handling novel concepts without explicit training examples. The model may struggle with accurately capturing the nuanced characteristics of these unfamiliar classes, leading to sub-optimal performance. Transfer learning, while beneficial for leveraging pre-existing knowledge, poses challenges in domain adaptation. Mismatches between the source and target domains can hinder the model's ability to effectively transfer learned features. Additionally, fine-tuning, while essential for task-specific refinement, demands careful consideration of hyperparameters to avoid overfitting or insufficient adaptation. The risk of catastrophic forgetting, where the model loses information from the original task during fine-tuning, is another challenge to address. Furthermore, balancing the trade-off between preserving valuable pre-trained knowledge and adapting to the specifics of a new task is a delicate aspect in these training approaches. Successfully addressing these challenges is pivotal for harnessing the full potential of zero-shot learning, transfer learning, and fine-tuning in enhancing model adaptability and task performance.

7   Conclusion and Future Work

Our comprehensive survey has shed light on the evolution, challenges, and versatile applications of prompt-based image generation models, laying the groundwork for future research endeavors. By addressing key aspects such as model specifications, user input guidelines, and training techniques, we have provided a holistic understanding of the current state-of-the-art. We wish that this paper serves as a stepping stone, inspiring further advancements and innovations in the dynamic intersection of language and vision within the realm of generative models. In the future, our research aims to delve deeper into the realm of prompt-based image generation models by conducting a comprehensive comparative analysis of the training techniques we discussed. This analysis will be instrumental in unraveling the nuances and intricacies associated with the training processes of these models. We envision exploring different training methodologies, optimization strategies, and fine-tuning approaches employed in the development of prompt-based image generation models. By systematically comparing the strengths, weaknesses, and performance outcomes of these techniques, we aim to contribute valuable insights into the optimization of model training for enhanced image synthesis. Additionally, we plan to investigate the impact of transfer learning, zero-shot learning, and fine-tuning on the efficacy of prompt-based models, unraveling their respective roles in model training. This future work holds the potential to provide a roadmap for refining training practices, ultimately advancing the capabilities and performance of prompt-based image generation models.

References

1. Esser, P., Rombach, R., Ommer, B.: Taming transformers for high-resolution image synthesis (2020)
2. Ramesh, A., et al.: Zero-shot text-to-image generation (2021)
3. Rombach, R., Blattmann, A., Lorenz, D., Esser, P., Ommer, B.: High-resolution image synthesis with latent diffusion models (2021)
4. Nichol, A., et al.: GLIDE: towards photorealistic image generation and editing with text-guided diffusion models (2021)
5. Raffel, C., et al.: Exploring the limits of transfer learning with a unified text-to-text transformer (2019)
6. Radford, A., et al.: Learning transferable visual models from natural language supervision (2021)
7. Alec, R., Jeffrey, W., Rewon, C., David, L., Dario, A., Ilya, S.: Language models are few-shot learners. In: Advances in Neural Information Processing Systems (NeurIPS) (2019)
8. Brock, A., Donahue, J., Simonyan, K.: Large scale GAN training for high fidelity natural image synthesis. In: International Conference on Learning Representations (ICLR) (2021)
9. Podell, D., et al.: SDXL: improving latent diffusion models for high-resolution image synthesis. ArXiv Preprint ArXiv:2307.01952 (2023)
10. Haoran, H., Yuqing, G., Zhengxu, W., Mingyu, C., Zhaoyang, W., Jingwen, X.: Masked visual-language pre-training for cross-modal understanding. ArXiv Preprint ArXiv:2110.13872 (2021)
11. Lu, J., Batra, D., Parikh, D., Lee, S.: VilBERT: pretraining task-agnostic visiolinguistic representations for vision-and-language tasks. In: Advances in Neural Information Processing Systems (NeurIPS) (2019)
12. Lingyun, Z., Cheng, X., Yiren, W., Jingwen, X.: CLIP-ViT: contrastive pre-training and vision transformer for efficient image retrieval. ArXiv Preprint ArXiv:2203.09480 (2022)
13. Bender, E.M., Gebru, T., McMillan-Major, A., Shmitchell, S.: : On the dangers of stochastic parrots: can language models be too big?. In: Proceedings of the 2021 ACM Conference on Fairness, Accountability, and Transparency (FAT*) (2021)
14. Cho, K., Van Merriënboer, B., Bahdanau, D., Bengio, Y.: On the properties of neural machine translation: encoder-decoder approaches. ArXiv Preprint ArXiv:1409.1259 (2014)

15. Gatys, L.A., Ecker, A.S., Bethge, M.: Image style transfer using convolutional neural networks. In: Proceedings of the IEEE Conference on Computer Vision and Pattern Recognition (CVPR) (2016)
16. Vaswani, A., et al.: Attention is all you need. In: Advances in Neural Information Processing Systems (NeurIPS) (2017)
17. Serban, I., Sordoni, A., Bengio, Y., Courville, A., Pineau, J. : Building end-to-end dialogue systems using generative hierarchical neural network models. In: Proceedings of the Thirtieth AAAI Conference on Artificial Intelligence (AAAI) (2016)
18. Radford, A., Narasimhan, K., Salimans, T., Sutskever, I.: Improving language understanding by generative pretraining. OpenAI (2018)
19. Devlin, J., Chang, M. W., Lee, K., Toutanova, K.: BERT: pre-training of deep bidirectional transformers for language understanding. ArXiv Preprint ArXiv:1810.04805 (2018)
20. Bahdanau, D., Cho, K., Bengio, Y.: Neural machine translation by jointly learning to align and translate. ArXiv Preprint ArXiv:1409.0473 (2014)
21. Yu, H., Lu, Y., Yiyuan, Z.: Speech-language integration in spoken language understanding. In: Proceedings of the 56th Annual Meeting of the Association for Computational Linguistics (ACL) (2018)
22. Pennington, J., Socher, R., Manning, C.D.: GloVe: global vectors for word representation. In: Proceedings of the 2014 Conference on Empirical Methods in Natural Language Processing (EMNLP) (2014)
23. Mnih, V., Heess, N., Graves, A.: Recurrent Models of Visual Attention. In: Advances in Neural Information Processing Systems (NeurIPS) (2014)
24. Zhang, H., et al.: StackGAN: text to photo-realistic image synthesis with stacked generative adversarial networks. In: IEEE International Conference on Computer Vision (ICCV) (2017)
25. Saharia, C., et al.: Photorealistic text-to-image diffusion models with deep language understanding (2023)
26. Johnson, W., Rickel, J., James, L.: Generative interactive virtual environment (GIVE). In: Proceedings of the 2000 International Conference on Intelligent User Interfaces (2000)
27. Lenat, D., Guha, R.: CYC project. In: Proceedings of the 1989 Workshop on Principles of Knowledge Representation and Reasoning (1989)
28. Terry, W.: SHRDLU. In: Proceedings of the Spring Joint Computer Conference (1972)
29. Schank, R.: A conceptual dependency parser for natural language. In: Proceedings of the 1969 Conference on Computational Linguistics (1969)
30. Ruiz, N., Li, Y., Jampani, V., Pritch, Y., Rubinstein, M., Aberman, K.: DreamBooth: fine tuning text-to-image diffusion models for subject-driven generation. ArXiv Preprint Arxiv:2208.12242 (2022)

Blockchain and Beyond – A Survey on Scalability Issues

Oliver Tominski(✉) and Martin Drebinger

FernUniversität in Hagen, Chair of Communication Networks,
Unistr. 27, 58084 Hagen, Germany
`oliver.tominski@studium.fernuni-hagen.de`

Abstract. Blockchains have gained significant popularity in certain areas over the past decade. The establishment of a distributed consensus among untrusted participants has set new standards and enabled protocols that were not possible before. The substantial critique of the Bitcoin network centers on its high energy consumption. However, its popularity has shown majorscalablity problems such as slow completion of transactions and the high transaction fees, particularly especially when transferring small amounts of money. After outlining the underlying structural reasons for these problems, we will briefly explain ideas for addressing them, using several projects as examples. Beyond that, we present an alternative concept utilizing Directed Acyclic Graphs (DAGs) which achieves a unified decentralized consensus in the network without using a traditional blockchain. The IOTA project has made significant progress with the so-called Tangle in version 2.0 and represents a promising approach for various application areas.

1   Introduction

This is a review of current blockchain technology, with a focus on the scalability issue. The article is structured into three sections. First, a brief synopsis of the Bitcoin blockchain, including its problems and possible solutions. Second, an introduction of the fundamental problem found in blockchain technology, followed by a collection of articles and whitepapers detailing potential solutions to the scalability issue. In the final section, a novel approach addressing the scalability problem is introduced, whereby a graph structure is employed as a ledger. The practical utilization of this graph is demonstrated using a specific implementation, for which relevant articles and whitepapers are available.

2   Brief History of Blockchain

A persistent challenge in open and decentralized P2P networks without trusted instances is the problem of free-riding, where some participants benefit from the resources and services provided by others without contributing themselves. To mitigate free-riding, research in the early 2000s focused on payment systems to

H. Unger and M. Schaible (Eds.): AUTSYS 2023, LNNS 1009, pp. 141–157, 2024.
https://doi.org/10.1007/978-3-031-61418-7_7

incentivize sharing and fairness. However, all of the introduced systems have experienced various attacks and are not resilient in an environment with malicious actors. Although the network was intended to be decentralized, the only solution was to centralize it in parts and create trusted entities.

2.1 Bitcoin – A (Partial) Success Story

The advent of Distributed Ledger Technology (DLT) has made decentralized payment systems possible. One widely recognized payment system within this category is Bitcoin [19], which was launched in 2009. Since its introduction, there have been no significant failures or bugs, and this is one reason why its market capitalization is around 0.5 trillion EUR in 2023 [4]. The development community has implemented various enhancements to the protocol, known as soft forks, as well as splits in the protocol to further improve its functionality (hard forks).[1] Its primary benefit is the capability to establish consensus in a distributed network without the need for a centralized and trustworthy entity. This is achieved even when a significant fraction of the participants are acting maliciously, and is also effective in mitigating sybil attack [12], where attackers create numerous identities to manipulate the consensus finding process. The idea that bitcoin could establish itself as a widespread and independent means of payment has not yet become reality, and many believe that it never will for several reasons.

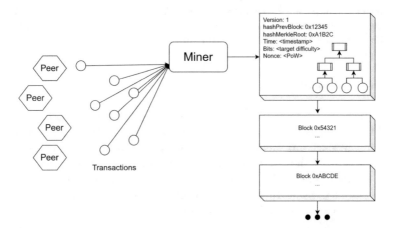

Fig. 1. Bitcoin protocol schematic

Brief Protocol Overview and Consensus Finding. To transfer cryptocurrency, a peer (user) must publish an unspent transaction output (UTXO) alongside a transaction fee. The UTXO [11] represents a cryptocurrency amount composed of previous transactions received. Miners collect transactions and group

---

[1] https://en.wikipedia.org/wiki/List_of_bitcoin_forks.

them into blocks (Fig. 1). Upon successful completion of a cryptographic brute-force challenge (Proof of Work PoW), a miner is granted authorization to append a block to the blockchain and receives a reward and additionally the transaction fees. The difficulty of the PoW challenge is adjusted that only every 10 min a block is added. This is to avoid frequent blockchain splits, as sufficient time is required for every peer to become aware of the newly added block. Since each block of the blockchain can only accommodate a limiting number of transactions (regardless of their amount), many transactions compete for their settlement. The higher the transaction fee provided, the higher the probability of prompt processing.

As the blockchain grows, there is a possibility that blocks could be added at the same time due to chance or malicious intent (i.e. fraud, double spending [8]), which could lead to a split in the blockchain. In such cases, multiple branches must be taken into account. The "longest chain wins" rule is utilized to establish consensus among peers, indicating that the longest chain has required the most effort (PoW) to construct. All transactions from abandoned branches are deemed invalid, and the involved miner will not receive any rewards or transaction fee.

The linearization of the blockchain serves three proposes, as stated by Bagaria et al. [1]. When a new block is appended, it selects the leading branch, appends new transactions to the ledger, and confirms the validity of all ancestor blocks in the selected chain.

Bitcoin – Technical Criticism

Power usage. The usage of PoW has resulted in the Bitcoin network's notable energy consumption. To gain the reward, miners must invest in specialized hardware and electric power to win the brute-force race. Using Proof of Work (PoW) can help mitigate the Sybil attack, provided that no coordinated group possesses over fifty percent of the mining power. Adjusting the difficulty level aids in preventing forks and ensures sufficient time for newly mined blocks to propagate. However, this also results in a consistently increasing need for computational power.

Throughput. Each block has a limited size of around 3000–4000 transactions (August 2023, [5]). Every 10 min a newblock emerges, resulting in a maximum throughput of around 7 transactions per second (TPS). This is many orders of magnitude less than the current established payment systems, e.g. the Visa credit card handles an average of 8100 TPS [29].

Transaction Fees. As the popularity of the Bitcoin network grows, there are an increasing number of transactions to be settled. Consequently, a transaction fee incentivizes miners to select transactions intotheir current block. This highly variable fee, which was around 2.46 USD end of April 2023 [2],is needed for a transaction to be included timely and makes transfers of small amounts unprofitable.

Storage. Regularly adding new blocks to the ledger results in a linear increase in storage requirements. Approximately 470 GB (June 2023 [3]) of storage capacity is required on each full-node within the network. This requirement contributes towards centralisation as fewer nodes are willing to devote their resources (computational power and storage) to participating in transaction validation and mining.

Consensus Finding. Establishing group consensus or global agreement on the current state of each member's assetspresents a significant challenge for all DLT or blockchain systems, particularly in a byzantine environment.This challenge is fundamentally described in a paper from 1985.

FLP Impossibility. The proof from **F**ischer, **L**ynch, and **P**aterson (FLP) which is called "FLP impossibility" shows that reaching a deterministic agreement in an asynchrone setting is impossible [13]. It is based on the "transaction commit problem" of distributed databases.

Fig. 2. schematic of FLP impossibility

It has been shown, that it is impossible to achieve the three objectives of termination (liveness), agreement (finality), and validity (integrity, fault tolerance) simultaneously (Fig. 2). Termination is the process whereby all honest nodes arrive at a decision and the system does not stall. While agreement denotes the network's capacity to reach a definite consensus or final commitment and validity implies fault tolerance against system entities' dropouts.

It is essential to achieve all three objectives through a blockchain consensus, but how is this practically achieved? One of the important goals is to achieve fault tolerance, which is intended to provide resistance in a Byzantine environment. It provides robustness and is achieved by using methods such as Proof of Work (PoW), resulting in protection against Sybil attacks. The other chosen objective is liveness, which involves the blockchain system establishing an uninterrupted process of consensus finding, such as through selection (longest chain) or a voting mechanism. The pursuit of safety or final agreement is jeopardized by a high likelihood of a final commitment only after a certain time. In the case of Bitcoin, this occurs after the completion of around six blocks. (0.1% after 6 confirmations [17]).

Byzantine Protection: Proof of Stake PoS. To address the problem of excessive energy consumption caused by the proof-of-work protocol, in 2012 King et al. [14] introduced an alternative solution called proof-of-stake (PoS) as a complement to bitcoin. In general, block validation is typically carried out by a single entity, chosen from a pool of candidates. The validation process is not computationally expensive and includes a reward (transaction fees) for the validator. To become a candidate, one must lock up (stake) an amount of their own currency. The validator selection criteria depend on the implementation, but generally include the amount staked (trust), time spent staking (reliability), and a random factor (fairness). To guarantee that validators perform their duties correctly, they may be penalised (slashed) by forfeiting a portion of their stake. This could also be initiated by other network entities (fact-check) subsequent to the attachment of a verified block to guarantee that the validator refrains from fraudulent behaviour. Not only does slashing serve as an incentive for validators, but a lack of diligent work from them undermines trust in the currency system as a whole, contributing to a decline of value in comparison to other currencies. This is of particular concern to entities holding significant value (high stakes).

Comparision. Using Proof of Stake (PoS) instead of Proof of Work (PoW) can significantly reduce energy consumption and eliminate the need for specialized mining hardware. It also reduces tendencies for centralization caused by mining pools (accumulation of miners). However, it is important to note that in order to earn money (become a validator), one must have coins for staking. With Proof of Work (PoW), individuals can win the competition through luck alone and create value without preconditions. This sets PoW apart from Proof of Stake (PoS), which requires values to already be available. Consequently, PoW is a more democratic approach than PoS.

3   Optimizations Regarding Scalability

In the next section, we explain why optimizing Bitcoin for scalability is not trivially possible and which approaches have nevertheless been found in literature.

3.1   Approaches to Mitigate the Scalability Limitations of Traditional Blockchains

Sharding. The research article by Croman et al. [10] highlights scalability as the principal difficulty facing the Bitcoin system (and all blockchains of this kind). The authors propose altering the blockchain (layer 1) itself as a possible solution, such as through the use of sharding to partition and distribute the blockchain among multiple hosts. However, reaching a consensus among multiple shards becomes challenging given the context of untrusted hosts, particularly in a byzantine fault environment.

Block Size and Interval. An alternative approach for a layer 1 solution would be to increase the block size or reduce the block intervals. This would allow for a greater number of transactions to be validated and settled per time, increasing throughput and decreasing transaction fees. However, this simple solution may unintentionally promote centralization. Reducing block intervals would decrease the amount of time for block propagation. Thus, the most recent miner who appended the last block and those miners with fast network connections towards them can initiate calculations for the next block earlier. So it is more beneficial to create mining pools. Increasing the block size results in greater storage usage for the entire ledger, thus raising the barrier for nodes to attain validator status as a full-node. Reducing fees results in an increase of smaller transactions that may bloat the blockchain due to transaction spam, ultimately leading to increase the potential of denial of service attack. Finally, both optimizations require a protocol upgrade (hard fork) for their implementation, which necessitates the approval of the network majority, an outcome unlikely to occur.

SegWit. Reducing the storage size of a Bitcoin transaction has been realized by a soft fork called Segregated Witness (SegWit [16]). By separating transaction signatures from their data, the block's capacity is effectively increased without increasing its size.

3.2   Blockchain Trilemma

A broader perspective on the blockchain optimisation problem, known as the blockchain trilemma, is presented in the following section.

Xiao et al. [32] have identified a trilemma between the objectives of security, decentralization, and scalability in blockchain consensus protocols. Only two out of three desirable properties can be achieved simultaneously. According to Xiao et al. [32], security is the paramount concern for blockchains in the financial domain. In this context, security pertains to consensus security within a Byzantine environment with malicious entities. This corresponds to the safety and liveness properties in classical distributed consensus (Fig. 3).

Reducing decentralisation leads to a more dominant central entities, which other network parties will be forced to trust. However, the inclusion of (trusted) central components undermines the key purpose of blockchain, which is to function in a decentralized Byzantine setting without requiring trust from any party. And thus, there exist more effective and efficient methods to accomplish the blockchain task with reliable components.

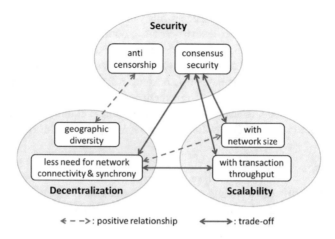

Fig. 3. Blockchain Trilemma (source [32])

Scalability is therefore frequently compromised. Xiao et al. give the example of Bitcoin where the security of Bitcoin hinges on long block intervals and best-effort mining to prevent double-spending, which results in low transaction throughput.

3.3   Layer 2 Solutions

Offloading work to higher layers is a strategy to boost transaction throughput in blockchains. Layer 2 is situated above the base layer, which provides consensus and transaction settlements, and can be utilised to outsource tasks without requiring any programmatic changes to the main blockchain (layer 1).

In the following, we would like to present a few selected approaches that address the scalability problem. The concepts are not always clearly distinguishable from one another and overlap to some extent or implement similar solutions in slightly different ways.

Smart Contracts. To extend the utilization of blockchain beyond its function as a financial ledger and take advantage of layer 2 protocol advancements, the majority of currently operating blockchains have incorporated a smart contract feature. Smart contracts are defined as "decentralized agreements built in computer code and stored on a blockchain". The notion was introduced by Nick Szabo in 1994 [26] to formalize and secure relationships over computer networks. This term was later adopted in the blockchain sector and the concept became the basis for the well-known Ethereum blockchain [7] which provides a decentralized Turing-complete virtual machine. A smart contract is an application that operates based on specific conditions being met and does not require a trusted intermediary. The code becomes immutable and transparent once it is deployed (and opensourced). Due to the irreversibility of the application, it is advisable to mathematically verify the correctness of the code prior to deployment. Smart contracts and more sophisticated applications, called decentralized applications (dApps), have the ability to establish their own arbitrary rules for ownership, transaction formats, state transition functions and are self-executing. For running the applications a reward for the executor is necessary. For instance, in the Ethereum blockchain, the execution process incurs a monetary cost known as "gas", which compensates the miners or verifiers for their efforts.

3.4   Payment Channel

Smart contracts provide a means of addressing scalability issues. Payment channels present one potential solution by facilitating the exchange of money amongst a group of entities (which trust each other). When the entities establish a payment channel on the primary blockchain, they can transfer funds between themselves. Once all exchange transactions are complete, the final outcome is updated by adjusting all account balances back onto the primary blockchain.

As an example, Bitcoin has implemented smart contracts, known as scripts [31], since around 2020 to allow the creation of payment channels. A user A can open a payment channel to another (known) user B by registering the connection in the blockchain (on-chain) and depositing a certain amount. Small payments, known as micropayments, can now be transferred between users A and B up to the deposited amount. Bidirectional channels can also be established. Multisigned transactions maintain mutual account balances, with each transaction including a timestamp. In the event of a dispute or settlement, the most recent transaction is used to settle the relevant balances on the main Bitcoin blockchain. Only two operations (transactions) on the Bitcoin main chain are required in total (opening and closing).

Bitcoin Lightning Network. Connecting multiple payment channels creates the Bitcoin Lightning Network[23], which became popular around 2020. It connects the bidirectional payment channels together and so transfers of funds are possible between users without a direct payment channel. A fully meshed network is avoided by publishing routes via a gossip protocol [6]. The incentive for routing isa small fee which is given for each hop. A sender must estimate the number of hops to reach a receiver and adjust the given fee accordingly.

Downfall of the Lightning Network. The work of Lin et al. [23] revealed that the network suffers from a centralisation bias. Due to its design, nodes tend to act as a central hub to make routing decisions and collect as many transaction fees as possible to maximise their profit.

3.5   Side Chains

Separated blockchains that are connected to a main blockchain are referred to as sidechains (Fig. 4). The connection allows data and assets to be exchanged

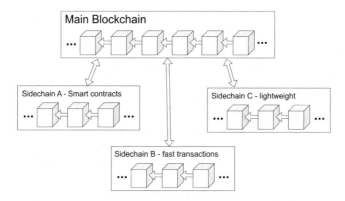

Fig. 4. sidechain schematic (based on *What is sidechain?* [30]).

(cross-chain transfer protocols CDT). This mechanism ensures that assets (such as cryptocurrency tokens) can be locked and deposited on the main blockchain and then be used on the sidechain, and vice versa. The main blockchain can offload work by using sidechains, which are often more centralized but may offer faster transactions and shorter confirmation time. This is possible through different consensus protocols, like Proof-of-stake, Proof-of-identity, and block parameters. The flexibility allows sidechains to optimize for specific use cases like smart contracts, general-purpose applications (dApps), identity management, supply chain tracking and so on. However, this does lead to a tradeoff between security (main) and speed (sidechain).

3.6 Optimistic Rollup

Using a sidechain to enhance scalability is achieved through a technique called optimistic rollup [20]. Transactions are executed, bundled, and optimised separately from the primary chain, with only transaction data (the summarized outcome of transactions) added to the primary ledger in a single batch, greatly reducing the number of main chain transactions. In an optimistic rollup, it is assumed that all transactions are usually valid. Anyhow, in order to receive punishment if fraud by the submitting party is detected, a deposit must be locked up when submitting batches to layer 1. After the batch is submitted, a fraud-proof process is initiated during the so-called challenge period. If other users detect an invalid transaction, the suspicious transaction is unbundled and executed again on the main chain. If the transaction is found to be invalid, the transaction creator is punished. Conversely, if it is deemed valid, the challenger is punished.

4 Directed Acyclic Graph as DLT

As an alternative for a conventional linear blockchain (Fig. 5a), a directed acyclic graph (DAG) can be employed. Every transaction is represented by a node and linked to other nodes by directed edges. The nodes create a network of interconnected relationships, and the acyclic directed edges display the flow of funds in chronological sequence with a well-defined history (Fig. 5b). Consensus depends on implementation. Transactions are confirmed over time by subsequent transactions that reference them, thereby incrementally increasing their validity. New transactions can be added with knowledge of only the minimum number of parent nodes, requiring only partial knowledge of the complete DAG. This allows the ledger to expand in parallel, resulting in short confirmation times and high scalability. The challenges to be addressed for this concept are transaction validation and confirmation, achieving consensus and synchronisation and maintaining a consistent DAG.

We present the approaches made possible by DAGs using the specific example of IOTA Tangle 2.0. Most of this section is basedon a publication from Müller et al. [18], which describes the ideas and mechanisms of Tangle 2.0 in depth.

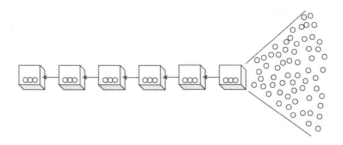

(a) Transactions in traditional blockchains

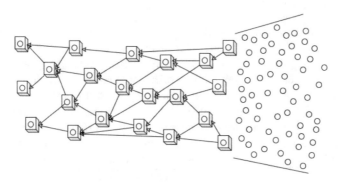

(b) Transactions in DAGs

Fig. 5. Transaction handling in different blockchain designs

4.1 Non-linear Structure

The term Tangle refers to the distributed data structure of the DAG, which represents the distributed ledger. Tangle 2.0 describes the design that is being developed under the code name Coordicide and, unlike its predecessor, is intended to work without a central authority. We will focus exclusively on this in the following. Every transaction in Tangle 2.0 is represented by a node and is linked to at least two previous nodes by directed edges. The nodes create a network of interconnected relationships, and the acyclic directed edges display the flow of funds in chronological sequence with a well-defined history. This is a significant distinction from traditional blockchains, where many transactions are bundled into a single block and these blocks themselves form the eponymous chain. The bundling of transactions requires the use of miners (context of PoW) or minters (context of PoS), which collect the transactions accordingly. This is not necessary in Tangle 2.0, as each participant can initiate their own transactions.

Consensus depends on the implementation. Transactions are confirmed over time by subsequent transactions that reference them, thereby incrementally increasing their validity. New transactions can be added with knowledge of only the minimum number of parent nodes, requiring only partial knowledge of the complete DAG. This allows the ledger to expand in parallel, resulting in short confirmation times and high scalability. The challenges to be addressed for this concept are transaction validation and confirmation, achieving consensus and synchronisation and maintaining a consistent DAG.

4.2 On-tangle Voting

A newly added block confirms its predecessor blocks and also their decisions about their predecessor blocks. Not all blocks have equal voting power due to the need to implement mechanisms such as proof of work or other similar methods for achieving byzantine fault tolerance (BFT). Popov et al. [25] describing a variant called Mana, which combines Proof of Stake and activity and is to be implemented in IOTA [24]. Voting weights continue to have an effect via the outgoing edges. A transaction becomes final when the weight assigned to it surpasses a specific threshold. Figure 6 outlines this approach: While block x and block y are not (yet) finalized, block z already is. Once a transaction is finalized, it becomes irrevocable.

Each block in Tangle 2.0 has a past cone and a future cone. The past cone comprises the transactions that can be directly accessed through the edges extending from the block or recursively through the outgoing edges of these blocks. The future cone includes blocks that refer to the considered block directly or indirectly (Fig. 7). In terms of voting, the past cone comprises blocks that are voted for, and the future cone contains blocks that vote for this block. The entire past cone of a finalized transaction is thus also finalized. For validation, it is therefore not necessary to run through the entire graph, but only up to the first completed blocks.

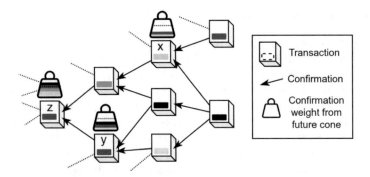

Fig. 6. On-tangle voting with weights, based on [18]

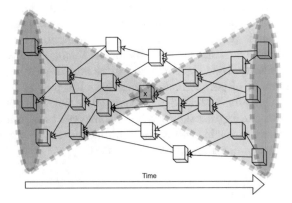

Fig. 7. Past and future cone of x, based on [18]

4.3   Conflict Resolution

A basic assumption of Tangle 2.0 is that the vast majority of transactions are valid. Nevertheless, the handling of invalid transactions must be ensured. In particular, double spending is a fundamental threat to any digital currency. This means that no more than one of two transactions transferring the same value can be validated [8].

A block that confirms contradictory transactions makes itself invalid. Honest participants will try not to vote for invalid or contradictory blocks. In order to generate a majority of votes for one of the two contradictory blocks and thus validate it and maintain liveness, there is the principle that new blocks from honest participants vote for the "heaviest" subgraph in this case. If this reaches a certain threshold of voting weights as described above, it is considered permanently confirmed or finalized. For the contradictory block, assuming that the necessary majority of participants are honest, there will no longer be a sufficient majority of votes to validate it. This block remains permanently unconfirmed.

With a threshold value of e.g. 60% required for finalization, this can easily lead to a deadlock if the voting power is largely evenly distributed between the two conflicting transactions. This deadlock would permanently block the finalization of one of the two subgraphs and thus endanger the entire Tangle, as an ever larger part would not be finalized from there on. To deal with this, it is possible for non-finalized blocks to subsequently change their voting decision. In this way, they can join the somewhat "heavier" subgraph and thus contribute to a clear election that exceeds the threshold value.

4.4   Side Effects and Their Treatment

If there is only one block with a contradictory transaction in a discarded subgraph, this is not a problem, but exactly the desired behavior to create a distributed consensus on an inherently valid structure. However, if there happen to be other valid and non-conflicting blocks in the subgraph, this would never be finalized, although it would actually be possible and sensible.

To deal with this, in addition to the normal relations (or edges) that we have looked at so far and which are used to vote for blocks, there are also so-called transaction votes. These do not contain the actual voting weight but merely "adopt" the transaction in the block concerned. In this way, transactions can be finalized and thus saved from an abandoned subgraph. It is important to note that the transaction votes relate exclusively to the linked block and are not inherited, as is the case with the voting weights of the block votes.

4.5   Bait-and-Switch Attack

In addition to the attacks that are always possible in distributed ledgers, in particular the so-called 51% attack[2], which ultimately corresponds to the takeover of the network by malicious participants, there are other process-specific attacks, we would like to briefly explain one example that is specific to this kind of distributed ledger and possible defense mechanisms. Conti et al. [9] present other security challenges in detail and in greater depth. The specific topic of parasite chain detection is described in [21], and routing attacks in [22].

There are various targets. Not only double spending, i.e. attacking the actual values in the ledger, can be a problem. For example, the functioning of the distributed ledger, the so-called liveness of the ledger, is also an asset worth protecting for honest participants, which must be defended against attackers. The attack that we want to exemplify here, because it specifically targets the peculiarities of the Tangle, is the bait-and-switch attack.

A malicious actor needs a larger amount of voting weight for this type of attack. The idea is that if two contradictory transactions are supported (voted) by one subgraph each, the actor always supports the slightly inferior subgraph. As soon as the latter moves towards a majority above the threshold, it changes

---

[2] Although the actual proportion of malicious participants required for this attack may be less than 51%, depending on the used consensus finding protocol.

its vote to the other one, causing other supporters to also change their vote to support the heavier subgraph and thus reach a consensus. By constantly switching, the attacker prevents this consensus and the functionality of the ledger is restricted as the following transactions cannot be confirmed.

As a countermeasure, a method called SRRS is proposed in [18]. According to this, the peers are only allowed to change their choice once in a certain (synchronized) time span and furthermore use a distributed random number generator to determine the subgraph to be chosen. In [15], the procedure is successfully simulated and ensures the liveness of the DAG up to an attacker weight of 33%.

5   Conclusion

Because DAG-based solutions do not require a single global view to be generated by the distributed ledger, the limitations imposed by the blockchain dilemma can be mitigated. These solutions have the potential to offer very low transaction fees while avoiding centralization by miners or large mining pools gaining excessive power. Additionally, some of the other disadvantages found in Bitcoin, such as the enormous amount of energy consumption or the need for specialized mining hardware, are not present in this case. Furthermore, it is assumed that the system scales very well and is fundamentally suitable for processing large volumes of transactions. This makes completely new use cases possible that were inconceivable with previous cryptocurrencies due to their limitations.

The biggest disadvantage is probably that the current implementations, including IOTA, are still under development. There can be no talk of a battle-tested solution like Bitcoin here. Quite the opposite: in the past there have already been serious security vulnerabilities, e.g. the trinity attack [28]. A single functioning and fundamental attack can destroy the entire system forever.

But even if everything works as planned, challenges remain. The low transaction costs are desirable, but can easily lead to problems such as "blockchain bloat" and transaction spam. And even if one should rather speak of "ledger bloat" in the case of DAGs, the consequences remain the same. The resources required to store the ledger and also the validation operations on the ledger become increasingly complex and therefore expensive. The advantages are then bought at a high price.

Further problems arise from the structures in which new DLTs develop in practice. The tokens are often not created over time, as is the case with Bitcoin and others, but are initially fully available. Many of the initiators act as start-ups and initially finance themselves with so-called Initial Coin Offerings (ICO) to venture capitalists and keep the other part of the tokens for themselves. The IOTA foundation, the organization behind the Tangle 2.0 holds a large proportion of the tokens, which leads to a centralization of power in connection with proof-of-stake. According to [27], the 10 largest addresses hold 74% in November 2023. This is a massive concentration of voting power for the future Tangle 2.0 network. This is not inherent in the system, but it is the current reality.

Although emerging and developing cryptocurrencies are just considered as risky investments by the media, significant progress is being made behind the scenes especially in a technical view. Even if it receives little attention, development is advancing and cleverly applied algorithms and ideas are addressing well-known drawbacks. Nevertheless, much remains hidden in Bitcoin's shadow. It is uncertain which of these options will thrive in the long run and which will quickly fade away.

If cryptocurrencies are to serve a purpose beyond speculative investment, it is essential to analyze the known disadvantages of the most commonly used approaches, which are also the most prevalent in terms of market capitalization, and implement improvements. The future will determine whether these existing methods can be improved through optimization or if entirely new approaches are required. In any case, research will continue in this area because it raises interesting and fundamental questions in distributed systems without a central trust position.

References

1. Bagaria, V., Kannan, S., Tse, D., Fanti, G., Viswanath, P.: Prism: deconstructing the blockchain to approach physical limits. In: Proceedings of the 2019 ACM SIGSAC Conference on Computer and Communications Security, pp. 585–602 (2019)
2. Bitcoin Fees Per Transaction (USD). https://web.archive.org/web/202306291520 28/https://www.blockchain.com/explorer/charts/fees-USD-per-transaction. Accessed 29 June 2023
3. Bitcoin Fees Per Transaction (USD). https://web.archive.org/web/202307251524 39/https://www.statista.com/statistics/647523/worldwide-bitcoin-blockchain-size/. Accessed 25 July 2023
4. Bitcoin Market Capitalization (USD). https://web.archive.org/web/20230326162 410/https://ycharts.com/indicators/bitcoin_market_cap. Accessed 26 Mar 2023
5. Blockchain.com, Chart - Average Transactions Per Block.https://web.archive. org/web/20231116092327/https://www.blockchain.com/explorer/charts/n-transactions-per-block. Accessed 16 Nov 2023
6. BOLT 7: P2P Node and Channel Discovery. https://web.archive.org/web/ 20230207072529/https://github.com/lightning/bolts/blob/master/07-routing-gossip.md. Accessed 07 Feb 2023
7. Buterin, V. Ethereum: a next-generation smart contract and decentralized application platform (2014). https://web.archive.org/web/20230514074259/https:// ethereum.org/669c9e2e2027310b6b3cdce6e1c52962/Ethereum_Whitepaper_-Buterin_2014.pdf. Accessed 14 May 2023
8. Chohan, U. The Double Spending Problem and Cryptocurrencies(2021)
9. Conti, M., Kumar, G., Nerurkar, P., Saha, R., Vigneri, L.: A survey on security challenges and solutions in the IOTA. J. Netw. Comput. Appl. **203**, 103383 (2022)
10. Croman, K., et al.: On scaling decentralized blockchains. In: Clark, J., Meiklejohn, S., Ryan, P.Y.A., Wallach, D., Brenner, M., Rohloff, K. (eds.) FC 2016. LNCS, vol. 9604, pp. 106–125. Springer, Heidelberg (2016). https://doi.org/10.1007/978-3-662-53357-4_8

11. Delgado-Segura, S., Pérez-Solà, C., Navarro-Arribas, G., Herrera-Joancomartí, J.: Analysis of the bitcoin UTXO set. In: Zohar, A., et al. (eds.) FC 2018. LNCS, vol. 10958, pp. 78–91. Springer, Heidelberg (2019). https://doi.org/10.1007/978-3-662-58820-8_6

12. Douceur, J.R.: The sybil attack. In: Druschel, P., Kaashoek, F., Rowstron, A. (eds.) IPTPS 2002. LNCS, vol. 2429, pp. 251–260. Springer, Heidelberg (2002). https://doi.org/10.1007/3-540-45748-8_24

13. Fischer, M., Lynch, N., Paterson, M.: Impossibility of distributed consensus with one faulty process. J. ACM **32**, 374–382 (1985)

14. King, S., Nadal, S.: PPCoin: peer-to-peer crypto-currency with proof-of-stake (2012)

15. Lin, B., Dziubałtowska, D., Macek, P., Penzkofer, A., Müller, S.: Robustness of the tangle 2.0 consensus. In: Hyytiä, E., Kavitha, V. (eds.) VALUETOOLS 2022. LNICS, Social Informatics and Telecommunications Engineering, vol. 482, pp. 259–276. Springer, Cham (2023). https://doi.org/10.1007/978-3-031-31234-2_16

16. Lombrozo, E., Lau, J., Wuille, P.: Segregated Witness (Consensus layer). https://web.archive.org/web/20230526235457/https://github.com/bitcoin/bips/blob/master/bip-0141.mediawiki. Accessed 26 May 2023

17. Mechkaroska, D., Dimitrova, V., Popovska-Mitrovikj, A.: Analysis of the possibilities for improvement of blockchain technology. In: 2018 26th Telecommunications Forum (TELFOR), pp. 1–4 (2018)

18. Müller, S., Penzkofer, A., Polyanskii, N., Theis, J., Sanders, W., Moog, H.: Tangle 2.0 leaderless Nakamoto consensus on the heaviest DAG. IEEE Access. **10**, 105807–105842 (2022)

19. Nakamoto, S.: Bitcoin: a peer-to-peer electronic cash system (2008). https://bitcoin.org/bitcoin.pdf

20. Optimistic Rollups. https://web.archive.org/web/20231115134020/https://ethereum.org/en/developers/docs/scaling/optimistic-rollups/. Accessed 15 Nov 2023

21. Penzkofer, A., Kusmierz, B., Capossele, A., Sanders, W., Saa, O.: Parasite Chain Detection in the IOTA Protocol. (arXiv,2020,4)

22. Perazzo, P., Arena, A., Dini, G.: An analysis of routing attacks against IOTA cryptocurrency. In: 2020 IEEE International Conference on Blockchain (Blockchain), pp. 517–524 (2020)

23. Poon, J., Dryja, T.: The bitcoin lightning network: scalable off-chain instant payments (2016). https://web.archive.org/web/20231115111137/https://lightning.network/lightning-network-paper.pdf. Accessed 15 Nov 2023

24. Sanders, W.: Explaining Mana in IOTA (2020). https://medium.com/iotatangle/explaining-mana-in-iota-6f636690b916. Accessed 11 Oct 2023

25. Popov, S., et al.: The Coordicide. In: IOTA Foundation (2020)

26. Szabo, N.: Formalizing and securing relationships on public networks. First Monday. **2** (1997). https://firstmonday.org/ojs/index.php/fm/article/view/548

27. Top 100 IOTA Rich Address List. https://www.coincarp.com/currencies/iota/richlist/

28. Trinity Attack Incident Part 1: Summary and next Steps (2020). http://blog.iota.org/trinity-attack-incident-part-1-summary-and-next-steps-8c7ccc4d81e8

29. Visa Transaction Facts. https://web.archive.org/web/20231002045449/, https://usa.visa.com/solutions/crypto/rethink-digital-transactions-with-account-abstraction.html. Accessed 02 Oct 2023

30. What Is a Sidechain? — Horizen Academy (2023). https://main.horizen-academy-v2.pages.dev/academy/sidechains/

31. Wuille, P., Nick, J., Towns, A.: Taproot: SegWit version 1 spending rules (2020). https://web.archive.org/web/20230225081223/https://github.com/bitcoin/bips/blob/master/bip-0341.mediawiki. Accessed 25 Feb 2023
32. Xiao, Y., Zhang, N., Lou, W., Hou, Y.: A survey of distributed consensus protocols for blockchain networks. IEEE Commun. Surv. Tutorials. **22**, 1432–1465 (2020)

Emotion-Aware Chatbots: Understanding, Reacting and Adapting to Human Emotions in Text Conversations

Philip Kossack$^{(\boxtimes)}$ and Herwig Unger

FernUniversität in Hagen, Chair of Communication Networks,
Universitätsstraße 47, 58097 Hagen, Germany
philip.kossack@studium.fernuni-hagen.de

Abstract. Chatbots and virtual assistants have revolutionized human-computer interaction. As these digital entities become integral to our lives, there's a pressing need to infuse them with emotional awareness. Emotion-aware chatbots, capable of understanding and responding to human emotions and thereby adapting their own personality, hold the promise of enhancing user experiences across various domains. However, the research in this field is scattered across disciplines, making it challenging to navigate. This survey consolidates diverse research findings, providing an overview of emotion-aware chatbots' current state and future directions. It covers foundational concepts, the significance of emotional intelligence, various AI techniques, research methodology, and approaches to emotional awareness. This survey sets the stage for future research in this exciting field, fostering the development of more emotionally intelligent chatbot systems.

Abbreviations

ADD Attention Deficit Disorder
AI Artificial Intelligence
AIE Ambient Intelligent Environment
AIoT Artificial Intelligence of Things
APA Animated Pedagogical Agent
BDI Belief-Desire-Intention
CAI Computer-Aided Instruction
EEG Electroencephalogram
FOSP Flexible Online Learning Support Platform
HRI Human-Robot Interaction
NLP Natural Language Processing

1 Introduction

In the contemporary landscape of human-computer interaction, the emergence of chatbots and virtual assistants has brought about a profound shift in the way

H. Unger and M. Schaible (Eds.): AUTSYS 2023, LNNS 1009, pp. 158–175, 2024.
https://doi.org/10.1007/978-3-031-61418-7_8

we communicate with technology. These artificial agents have become an integral part of our daily lives, offering assistance, information, and entertainment. However, as their capabilities have expanded, so too needs to imbue them with a crucial element of human interaction: emotional awareness.

The ability of chatbots to recognize, interpret, and respond to human emotions represents a critical advancement in the field of artificial intelligence (AI). It holds the promise of creating more empathetic, effective, and engaging interactions between humans and machines. Whether it's providing emotional support to individuals struggling with mental health issues, offering guidance in customer service scenarios, or simply enhancing the overall user experience, emotion-aware chatbots have the potential to revolutionize the way we interact with technology.

Yet, despite the growing recognition of the importance of emotion-aware chatbots, the landscape of research in this domain remains scattered across various disciplines. Solutions and innovations are dispersed among fields such as natural language processing, affective computing, psychology, and machine learning, making it challenging for researchers, developers, and practitioners to navigate this complex terrain.

This survey aims to provide a comprehensive overview of the state of the art in emotion-aware chatbots by consolidating and synthesizing research findings from diverse fields. By drawing insights from these disparate sources, this work seeks to shed light on the current state of the field, identify common themes and challenges, and chart a course for future research and development in this exciting and transformative area of AI.

It begins by presenting a general chatbot in Sect. 1.1, followed by an explanation of the importance of emotionally intelligent chatbots in Sect. 1.2.

This paper continues with a description of the different aspects of emotion in artificial intelligence in general in Sect. 2. The research method is lined out in Sect. 3. The presentation of different approaches to emotional awareness and adaption is then given in Sect. 4. Finally, Sect. 5 offers a starting point for further research in this area.

1.1   Chatbots

[2] introduces a general architecture for chatbots. As shown in Fig. 1, a user enters a request in a user interface. This request is received by the chatbot. A Language Understanding Component infers the intent of the user and the provided context information. This extracted information is then forwarded to a Dialogue Management Component. This component holds the conversation context, calls the Action Execution Component and the Information Retrieval Component and passes the results to the Response Generation Component. The Information Retrieval Component is responsible for querying Data Sources, like a knowledge base or a web resource. The Action Execution Component is responsible for performing the requested action. The result of the action or the information retrieved is passed to the Response Generation Component together with the desired response. The Response Generation Component uses Natural Language Generation to generate the response based on the conversation's context, the action executed and the information retrieved.

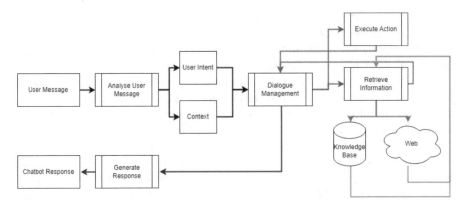

Fig. 1. General Chatbot Architecture according to [2]

The classification of chatbot models is based on how they process inputs and generate responses. Rule-based chatbots rely on predefined rules and patterns, retrieval-based chatbots fetch responses from a database, and generative chatbots use advanced machine learning techniques for more human-like interactions, although they can be more challenging to develop and train [5].

According to their primary objectives, chatbots can be classified into informative, conversational and task-based chatbots. Informative chatbots provide information, conversational chatbots engage in human-like conversations, and task-based chatbots are designed to perform specific tasks while intelligently handling user interactions related to those tasks. All of them can work either in the open domain, on general topics, or in the closed domain, only on specific topics [5].

Based on the service provided, chatbots can be classified into interpersonal chatbots, which act as intermediaries for various tasks, intrapersonal chatbots, which provide personalized companionship, and inter-agent chatbots, which facilitate communication between different chatbot entities [4].

1.2   The Case for Self-adapting Emotionally Intelligent Chatbots

Emotionally intelligent and empathic chatbots have the potential to transform various domains, from mental health support to customer service, and raise important ethical and human-computer interaction questions that warrant investigation. Therefore it is essential to understand their potential benefits, limitations, and societal implications.

Negative emotions experienced by users during interactions appear to be the primary factor leading to conversation abandonment. Conversely, positive emotions not only serve as a significant deterrent against communication breakdowns but also have the potential to enhance the results of a technological intervention [10].

The chatbot's 'mood' could also influence users' willingness to sustain the interaction. In [6], the authors examined how children responded to four distinct chatbots, each possessing different personalities and moods. Surprisingly, despite the absence of any aggressive behavior directed toward the ill-tempered chatbot, children spent the least amount of time engaging with it, demonstrating a preference for the other chatbot agents instead.

As well as specific emotions, a chatbot's empathy in general can change the interaction between humans and chatbots [10]. As an example, [7] illustrates that when users seek advice from a chatbot regarding a health issue, the chatbot's display of affective empathy is perceived as more supportive by users compared to merely offering medical information. In [8], the authors emphasize the emotional support that the chatbot can offer, encompassing expressions of care, affection, and empathy. Such expressions have the potential to alleviate users' feelings of loneliness.

It is important to note, that this finding doesn't apply to everyone. According to [9] some users with a strong expectation that a machine has to behave unemotionally feel uncomfortable and even annoyed when a chatbot pretends to possess emotions and can understand theirs.

So far, the focus of emotionally intelligent chatbots is to recognize and respond to the emotions of the users. For example, a chatbot that notices that its counterpart is sad can try to cheer them up or offer them help. A chatbot can also observe the effect of its reactions and learn from them how to improve its communication skills. What does not exist yet is a chatbot that has feelings itself and knows when to react. Such a chatbot would not only adapt to the emotions of its interlocutors but also be able to change its own emotions, depending on the situation and goal. For example, a recent analysis of the emotional capabilities of chat-gpt conducted by [43] concludes that chat-gpt can understand emotional texts and can generate emotional responses. The further criteria mentioned above are not considered.

By considering not only literature about chatbots but also about agents in general from different research topics reaching from pedagogics to health care, this paper presents the overall state of the art in systems that have their own emotions, understand emotions, react according to their own emotions and the emotions understood and can adapt themselves according to the conclusion drawn from changes in own and external emotions.

2   Emotion in AI

Section 2.1 gives an overview of emotion in artificial intelligence. Learning based on emotion is described in Sect. 2.2.

2.1   Overview

In [12] an overview of emotion in artificial intelligence is provided. The authors identify three major domains, which are affective computing, emotion as reinforcement and anatomical design.

Affective computing is defined as a domain dedicated to equipping computer systems with the resources to recognize, express and even have emotions. In emotion recognition, user behavior is mapped to an emotional state in a categorical (happy, sad,...) or dimensional (arousal-valence and sometimes dominance) emotional space. See [14] for further information on emotional models. There are approaches for different modalities. Modalities considered are facial expressions, speech, text and biological signals [15]. As in many fields of text-based NLP, text-based emotion recognition approaches can be divided into rule-based and machine-learning-based methods. State of art performance is usually achieved with the use of large pre-trained language models. See [13] for a survey on text-based emotion classification. Emotion as reinforcement uses emotions as reward and conditioning in the cognitive development of agents. Agents, with no natural internal drives, can get motivation from emotion which can improve efficiency and exploration. In the anatomical design domain, emotional concepts are implemented as circuits inspired by the design of the brain, especially the limbic system.

2.2   Learning Based on Emotion

[17] presents a biologically inspired decision-making system for autonomous agents, both physical and virtual. As shown in Fig. 2, the system is based on drives, motivations, and emotions. Each artificial emotion is treated separately, and the generation method and role of each emotion are not defined as a whole. Instead, the agent can learn the emotional releaser and actions based on its own experience. The proposed system has been implemented on virtual agents living in a simple virtual environment. The experiments show that the learning process produces natural results, and the artificial emotions significantly improve the performance of the agent.

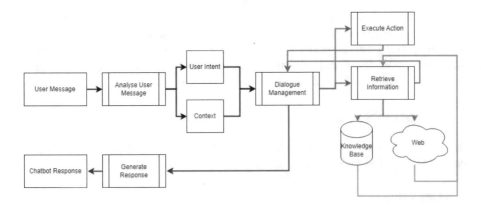

Fig. 2. Learning based on emotion according to [17]

In [18] a generic model for personality, mood, and emotion simulation for conversational virtual humans is presented. The model is designed to update the parameters related to emotional behavior and is based on Bayesian networks. The paper explores how existing theories for appraisal can be integrated into the framework. The proposed model helps to construct more authentic cognitive behavior models and simulate irrational behavior and fully self-organized collective behaviors of human populations.

The paper describes a prototype system that uses the described models in combination with a dialogue system and a talking head with synchronized speech and facial expressions. The system is capable of diagnosing the emotions and personality of the user and generating appropriate behavior by an automated agent in response to the user's input. The proposed model has been tested on various conversational agents, and the results show that it can generate more natural and engaging conversations with users.

3 Methodology

In this section, the methodology employed for this survey is described. The approach for identifying appropriate candidates for the review is outlined in Sect. 3.1. The criteria dedicated to comparing and analyzing various approaches are detailed in Sect. 3.2.

3.1 Search Method

The survey is conducted based on the approach of Webster and Watson [19].

To identify suitable candidates for inclusion in the review, a comprehensive search was conducted using a combination of search terms. Where possible, the search was restricted to the fields Title and Abstract. The search terms used were "emotion* AND *agent AND adapt*", "emotion* AND chatbot AND adapt*", "emotion* AND chat-bot AND adapt*", "emotion* AND *assistant AND adapt*". These search terms were chosen to ensure broad coverage of relevant literature in the field.

The search was performed in IEEE Explore (https://ieeexplore.ieee.org/) and dblp (https://dblp.org/). These databases were selected due to their extensive collections of research papers in the area of interest.

Initially, a search term-based database search was conducted to retrieve a set of candidate papers. Subsequently, a manual analysis of the abstracts was carried out to determine whether each candidate paper was within the scope of the review. The remaining papers were reviewed. This manual analysis helped to ensure the relevance and appropriateness of the selected papers.

For all papers that met the inclusion criteria, an additional step was taken to expand the candidate set. Specifically, the papers where included that cited at least one of the selected papers. This iterative process of identifying and adding relevant papers was repeated until no new candidates were found, ensuring comprehensive coverage of the literature.

3.2 Categorization of Approaches

To facilitate the analysis and comparison of the identified approaches, a set of aspects was defined for categorizing the approaches discussed in the selected papers. These aspects provide a structured framework for evaluating and understanding the different methodologies and techniques employed in the field of chatbots and conversational agents with a focus on adaptability and emotion.

The following aspects were considered for categorizing the approaches:

- Adaption of agent's emotion based on user's emotion
- Response selection based on user's emotion
- Learning based on emotion

4 Results

In total 25 papers met inclusion criteria and were fully reviewed. The distribution to different topics is as follows: The most frequent subject was pedagogical, with seven papers that addressed topics such as learning outcomes, instructional design, and assessment methods about emotional adaption. The second most frequent subject was health care, with four papers that explored the use of emotional adaption to improve the quality, accessibility, and efficiency of healthcare services. The third most frequent subject was agent, with three papers that focused on the use of emotional adaption in the development of intelligent agents in various domains. The fourth most frequent subject was human-robot interaction, with two papers that investigated the aspects of interacting with emotionally aware robots. The fifth and sixth most frequent subjects were chatbot and ambient intelligence, with one paper each that discussed the challenges and opportunities of designing and deploying emotional adaptive conversational agents and a framework for creating smart environments that can adapt to human needs and preferences based on their emotions respectively. Another seven of the reviewed papers proved as out of scope during the review.

Concerning the categories given above. The distribution of the papers to the categories and topics is described in Fig. 3. Two papers focus on the adaption of an agent's emotion based on the user's emotion. Nine papers deal with response selection based on the user's emotion and seven papers are about learning based on emotion.

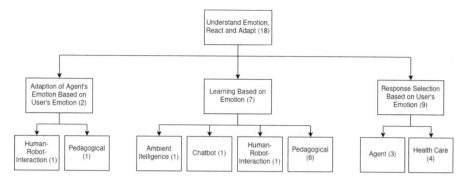

Fig. 3. Feature Categorization

The different methods used in the different approaches are given in Tab. 1. In total, five approaches make use of external data sources, three use fuzzy learning, four use neural networks, three use reinforcement learning and three are rule-based.

Table 1. Methods to Categories Matrix

Approach	Category		
	Adaption	Learning	Response Selection
External Data Sources	0	0	5
Fuzzy Learning	0	3	0
Neural Networks	1	0	3
Reinforcement Learning	0	3	0
Rule-Based	1	1	1

4.1 Adaption of Agent's Emotion Based on User's Emotion

Approaches in this section have in common, that the emotion of a user is used as a signal to adapt the agent's emotion. From two approaches in total, one belongs to human-robot interaction and one to pedagogic. They are described in Sect. 4.1 and in Sect. 4.1.

Human-Robot Interaction. A focus on the naturalness and effectiveness of human-robot interaction is the basis in [35]. The authors undertook the development of an agent aimed at facilitating intelligent interactions between humans and robots. Their work involves the integration of a computational model of artificial emotions that possesses learning and self-adaptation capabilities. Human gestures serve as triggers for altering the robot's "emotional state," manifesting through various channels such as integrated visual media, environmental lighting, music, and adjustments in the robot's style of movement and behavior. The authors employed their emotional agent architecture as the foundation for building this agent, which operates as an "emotional activator," stimulating human creativity within their environment. Notably, their system eliminates the need for individuals to wear specialized on-body sensors, opting instead for a small wheeled robot equipped with onboard ultrasound sensors. Furthermore, they enhance its sensory input capabilities through the development of a camera-based sensor system.

Pedagogy [24] introduced an interactive multi-agent system for child emotion recognition. Their approach leverages both cognitive and non-cognitive factors to estimate a child's real-time emotional intensity. A social agent and an autonomous intelligent agent collaboratively employ an adaptation model based on the child's emotional intensity to dynamically adjust the game's status. Additionally, an animated pedagogical agent provides motivational support, facilitating user interaction. Notably, their results indicate a robust 82.5% accuracy in recognizing a child's emotions through affective gesture recognition, while the social agent's estimation of emotional intensity demonstrates a strong correlation with observer feedback, particularly in higher intensity levels.

4.2 Response Selection Based on User's Emotion

In this section are those approaches, which select their response based on the emotion of a user. Of the nine papers that belong to this section, one is in the category of ambient intelligence, chatbot, and human-robot interaction while the remaining six papers are in the pedagogical area. They are described in the order mentioned in Sects. 4.2, 4.2, 4.2 and 4.2.

Ambient Intelligence. Home environment adjustment according to a user's emotional state is performed in [38]. This paper presents research on an emotion-integrated Artificial Intelligence of Things (AIoT) smart home automation system. Unlike previous systems that focused on remote control, voice commands, or virtual assistants, this system combines real-world data for emotion recognition, sensing, learning, and decision-making. It enhances security, efficiency, and user experience in smart homes by using computer vision, environmental sensing, and emotion recognition. The system employs face recognition for door access control and utilizes Convolutional Neural Networks and support vector machine algorithms for emotion detection, adjusting ambient lighting to match the user's mood. The paper introduces the concept of "sensovisual," a fusion of sensor and visual methods for smart home automation.

Chatbot. Situation-aware emotion regulation of conversational agents with kinetic earables is investigated in [29]. The authors conducted a mixed-method study involving 280 users and 12 qualitative interviews to understand user expectations. Based on those findings, they developed an emotion regulator for a conversational agent on kinetic earables. This regulator dynamically adjusts the agent's conversation style, tone, and volume in response to users' emotional, environmental, social, and activity contexts. The experimental results consistently show that our emotion regulation mechanism significantly enhances user experiences compared to baseline conditions across various real-world settings.

Human-Robot Interaction. A framework for cooperation with humans in a social context is given in [37]. In this study, the authors address the critical aspect of human-robot interaction (HRI) within assistive robotics. They emphasize the importance of robots understanding not just language but also human emotions, especially when digital assistants fail to meet users' expectations, leading to emotional responses. To bridge this gap, the authors propose an emotion-based perception architecture for cooperative HRI. This system enables the robot to recognize human emotional states, fostering a more natural connection. Additionally, the authors suggest using measured emotions as a metric to grade HRIs, which allows the robot to adapt its behavior. Emotions are detected through vision and speech inputs processed by deep neural networks, ensuring a better HRI experience, especially in cases where users exhibit negative emotions.

Pedagogical [20] explored the growing prominence of Animated Pedagogical Agents (APAs) in a range of instructional applications spanning various learning domains. APAs are increasingly recognized for their potential to enhance the user learning experience, thereby elevating the overall effectiveness of courseware. Additionally, the authors underscore the significant influence of learner emotions on cognitive processes. In this paper, the authors delve into a pilot study centered on the implementation of courseware featuring an APA that dynamically adjusts to the learner's emotional state. To evaluate its efficacy within the target demographic, the APA within the courseware underwent testing employing a pre-test and post-test methodology. Results from the study revealed that incorporating the learner's preferred emotion into the APA within courseware yielded superior learning outcomes compared to an APA devoid of emotional adaptation.

The paper [21] proposes the utilization of an intelligent emotional pedagogical agent as a means to enhance the communication interface between users and e-learning systems, specifically tailored for students with attention deficit disorder (ADD). Their argument centers on the premise that emotional communication between users and e-learning systems can significantly enhance interaction and foster a more socially engaging learning environment, accommodating learners from diverse backgrounds.

In their paper, the authors provide a succinct review of pertinent literature, outline the distinctive learning requirements associated with attention deficit disorder (ADD), establish a connection between emotional processes and ADD, delineate the key learning and adaptation factors influencing the agent's behavior, present the architecture of their model, and detail a small-scale experiment involving higher education students, including those with ADD. The application's effectiveness is assessed, with comprehensive data analysis and statistical findings presented to support their research outcomes.

The paper [22] introduced an innovative approach to address a prevailing limitation in traditional intelligent Computer-Aided Instruction (CAI) systems, namely the insufficient consideration of emotional cognition. Their solution involves the development of an emotional pedagogical system grounded in artificial psychology and Agent theories. In this system, the emotional state of students is incorporated as a secondary index within a two-layer fuzzy comprehensive assessment model for evaluating the effectiveness of the learning process.

The inference engine within this framework utilizes the assessment results to generate final teaching strategies and suggestions, drawing from a synthesis of cognitive and emotional factors. What sets this approach apart is its emphasis on the role of the Agent, which not only aims to achieve teaching goals based on cognitive factors but also actively senses and adapts to the psychological states of the students. The resulting pedagogical system integrates a dual-source mechanism for assessing and facilitating interactions that encompass both emotional and cognitive aspects, ultimately simulating a "man-to-man" classroom teaching experience.

[23] have centered their research on the application of a multi-agent system within e-learning environments, primarily geared towards streamlining adaptation processes and optimizing the attainment of e-learning objectives. Recognizing the significance of addressing the unique characteristics and emotional intelligence of e-learners, the authors have coined their system as "eQ."

The eQ multi-agent system places a primary emphasis on the BDI (Belief-Desire-Intention) agent rational model, which serves as a fundamental framework for the implementation of their proposed adaptation strategy, known as the FOSP (Flexible Online Learning Support Platform) method. This approach is crafted to align e-learning experiences with the genuine needs and well-established preferences of e-learners, fostering a more tailored and effective educational journey in the digital realm.

[25] delve into a comprehensive exploration of emotional pedagogical agents within intelligent tutoring systems. Their research journey begins with the formalization of domain knowledge representation and user models. Subsequently, the paper introduces a formal model for emotional pedagogical agents and meticulously elucidates its functionalities.

This particular emotional pedagogical agent stands out for its ability to incorporate the emotional states of users throughout the dynamic generation of personalized learning units. This process draws from information gleaned from user models and domain knowledge. The ultimate aim is to enhance the self-adaptability of the system and elevate its pedagogical effectiveness by tailoring the learning experience to the emotional needs and preferences of the users.

The authors in [26] propose the development of an intelligent virtual agent grounded in the principles of the Social Regulatory Cycle. This virtual agent serves the role of a coach and is readily accessible through a mobile application.

This virtual coach is designed to adapt to the unique needs of individual human students, offering support and assistance based on their intentions, motivations, and emotional states.

The virtual coach leverages both objective and subjective user data to interpret the learning context of each user. Employing a synthetic character with human-like mannerisms, the virtual coach guides users by modulating their emotional states within the learning process. Initial evaluations conducted within an online learning environment demonstrated that users maintained consistent interest and engagement with these virtual coaches, highlighting the potential of this approach to address motivational challenges in long-distance learning.

4.3   Learning Based on Emotion

This section describes approaches that use emotion as signals in learning. Three of the seven papers described here in Sect. 4.3 belong to the category agent. The remaining four papers in Sect. 4.3 are in the health care domain. Three of them are focused on the emotions of elderly people. One of them deals with keeping depressed people happy.

Agent [30] introduces the significance of emotion in human intelligence, highlighting its crucial role in agent performance and adaptability. The paper presents an emotional agent architecture that employs reinforcement learning for decision-making. Notably, the reward mechanism for behavior derives from both the agent's emotions and external environmental factors. Experimental results demonstrate that this emotional agent exhibits a faster learning rate and greater adaptability in its behavior compared to its emotionless counterparts.

A system called ARTEMIS is proposed in [33] as a framework for designing and implementing artificial emotions in intelligent systems. ARTEMIS possesses the capability to generate human-like artificial emotions during its interactions with the environment, and the paper delves into the underlying mechanisms supporting this functionality. Additionally, the control system is equipped to retain a history of these artificial emotions, which are stored within an Agent Knowledge Graph. ARTEMIS leverages both current and historical emotions to adapt its decision-making and planning processes. To validate the concept, the authors implement a tangible software agent employing the ARTEMIS control system as a user assistant, responsible for executing user commands in a dynamic environment populated with other autonomous service agents. These interactions lead to the emergence of artificial emotions within the user assistant, which is documented in its Agent Knowledge Graph. The initial experiments demonstrate the feasibility of creating an autonomous user assistant with plausible artificial emotions using ARTEMIS. Furthermore, the results underscore the beneficial role of recorded emotions in enhancing planning and decision-making capabilities in intricate and ever-changing environments, surpassing the performance of emotionless counterparts.

In [32], an adaptive model inspired by human behavior, emphasizing the dynamic interplay of emotions and personality within agents, is introduced.

Unlike fixed personality traits, the agent's emotions and personality evolve over time and in response to interactions. Decision-making takes place within the agent's environment and among other agents. In this model, the agent's personality directly influences its decisions, while emotions indirectly impact these choices. Agents perceive emotions, leading to changes in their personalities based on these emotional experiences. The study investigates agent behavior through multi-agent simulations across various scenarios, examining resource exchange and consumption for survival. Additionally, the agent's personality and behavior adapt based on the personalities of other agents encountered. Results highlight the agent's adaptability within its environment, shaped by the surrounding community and environmental factors.

Health Care [39] presents a novel intelligent agent, the Health Adaptive Emotion Fuzzy Agent (HAOEFA), designed to address the growing challenge of providing adequate care and well-being support for the elderly population living at home. The study highlights the increasing expectations for well-being and life expectancy among the elderly, emphasizing that traditional individual medical monitoring will become unsustainable. To bridge this gap, the paper proposes leveraging Ambient Intelligent Environments (AIEs) and introduces a learning technique that combines emotion recognition and fuzzy logic-based adaptation. HAOEFA is described as an automated self-learning system capable of adapting to individual needs, including emotional preferences and changing behaviors. The research findings demonstrate that this innovative agent can effectively provide both monitoring and ambient environmental control, enhancing the well-being of individuals with limited physical or cognitive abilities, even with limited external resources.

In [40], the authors propose a computational model to understand the affective states of the elderly based on their daily activities, aiming to improve the capabilities of a humanoid agent residing on a smartphone platform. The model leverages recent advancements in human emotion recognition, utilizing visual, audio, and keyboard or touchpad stroke pattern signals. By integrating this understanding into the mobile agent, the authors aim to make it more human-like, enhancing its ability to provide specific and appropriate care to the elderly population. The initial knowledge regarding activity-affect associations is drawn from psychology and gerontology literature, and the model adapts this knowledge based on training signals, ultimately enabling the mobile agent to better cater to the emotional needs of the elderly.

[42] introduces a novel intelligent fuzzy agent equipped with an emotion-aware system, effectively transforming the living environment into a pseudo-robotic entity in which elderly individuals reside. This system is designed to offer a non-invasive, self-learning, intelligent control mechanism that continually adapts to individual requirements. The paper also delves into a comparative analysis of two distinct self-learning methodologies developed for this purpose. Empirical findings from experiments conducted at the Glamorgan Intelligent

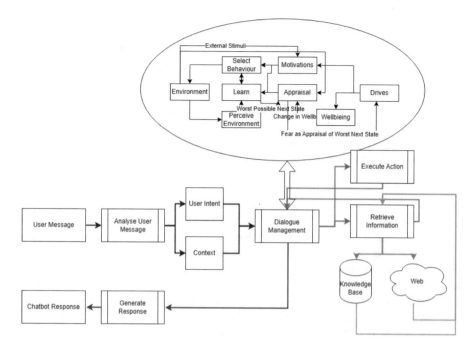

Fig. 4. Integration of Emotional Awareness in Chatbots

Home Care (Glam i-Home Care) are presented, showcasing the potential of this approach in facilitating the extension of independent living for seniors.

Music recommendation based on the user's emotional state is described in [41]. To address this issue, the study used EEG data to assess music's emotional impact and employed reinforcement learning to create personalized playlists aimed at enhancing emotional states. The research, based on data from 32 participants, successfully guided users towards joyful emotional states, with an average angular error of 57°. These findings have clinical relevance, offering potential improvements in music therapy for clinical depression and anxiety, particularly in remote and personalized settings where automated techniques can help select music to enhance emotional well-being.

5   Conclusion

This comprehensive survey has traversed the diverse landscape of emotion-aware chatbots, dissecting their multifaceted abilities to have their own emotions, comprehend human emotions, and tailor responses accordingly. While substantial progress has been made in individual facets of emotion awareness, a critical observation emerges from that exploration: the absence of an integrative app-roach that empowers chatbots with the holistic ability to experience, interpret, respond, and adapt in the emotional realm.

The survey has illuminated the fact that existing research often operates in isolated domains, addressing specific dimensions of emotion awareness. Chatbots may possess the capacity to simulate emotions, analyze user emotions, or even respond empathetically to user emotional cues. However, the full realization of emotionally intelligent chatbots - one that embodies its own emotional state, understands user emotions, responds authentically based on this understanding, and dynamically adapts itself in response to shifting emotional landscapes - remains a topic yet to be fully explored.

The future of emotion-aware chatbots lies in the pursuit of this integrative approach, as shown in Fig. 5. It calls for an interdisciplinary synthesis of insights from fields such as artificial intelligence, psychology, affective computing, and human-computer interaction. The development of such integrative systems is not without its challenges. Ethical considerations, privacy concerns, and the delicate balance between utility and intrusion must all be carefully navigated. Moreover, the creation of chatbots that evolve their emotional responses based on changing internal and external emotional states demands sophisticated algorithms, data streams, and machine learning models.

In conclusion, this survey underscores the need for future research to embark on the journey of constructing emotion-aware chatbots that encompass the com-plete spectrum of emotional intelligence. These agents, driven by their emotional state, equipped with the ability to fathom user emotions, respond authentically, and adapt to evolving emotional dynamics, hold immense promise in revolution-izing human-computer interactions.

References

1. Shum, H., He, X., Li, D.: From Eliza to XiaoIce: challenges and opportunities with social chatbots. Front. Inf. Technol. Electron. Eng. **19**, 10–26 (2018)
2. Adamopoulou, E., Moussiades, L.: An overview of chatbot technology. In: IFIP International Conference on Artificial Intelligence Applications and Innovations, pp. 373–383 (2020)
3. Sannon, S., Stoll, B., DiFranzo, D., Jung, M., Bazarova, N.: How personification and interactivity influence stress-related disclosures to conversational agents. In: Companion Of The 2018 ACM Conference On Computer Supported Cooperative Work And Social Computing, pp. 285–288 (2018)
4. Nimavat, K., Champaneria, T.: Chatbots: an overview types, architecture, tools and future possibilities. Int. J. Sci. Res. Dev. **5**, 1019–1024 (2017)
5. Hien, H., Cuong, P., Nam, L., Nhung, H., Thang, L.: Intelligent assistants in higher-education environments: the FIT-EBot, a chatbot for administrative and learning support. In: Proceedings of the 9th International Symposium on Information and Communication Technology, pp. 69–76 (2018)

6. Pérez-Marín, D., Pascual-Nieto, I.: An exploratory study on how children interact with pedagogic conversational agents. Behav. Inf. Technol. **32**, 955–964 (2013)
7. Liu, B., Sundar, S. Should machines express sympathy and empathy? Experiments with a health advice chatbot. Cyberpsychol. Behav. Soc. Netw. **21**, 625–636 (2018)
8. Ta, V., et al.: User experiences of social support from companion chatbots in everyday contexts: thematic analysis. J. Med. Internet Res. **22**, e16235 (2020)
9. Urakami, J., Moore, B., Sutthithatip, S., Park, S.: Users' perception of empathic expressions by an advanced intelligent system. In: Proceedings of the 7th International Conference On Human-Agent Interaction, pp. 11–18 (2019)
10. Rapp, A., Curti, L., Boldi, A.: The human side of human-chatbot interaction: a systematic literature review of ten years of research on text-based chatbots. Int. J. Hum.-Comput. Stud. **151**, 102630 (2021)
11. Catani, M., Dell'Acqua, F., De Schotten, M.: A revised limbic system model for memory, emotion and behaviour. Neurosci. Biobehav. Rev. **37**, 1724–1737 (2013)
12. Assunção, G., Patrão, B., Castelo-Branco, M., Menezes, P.: An overview of emotion in artificial intelligence. IEEE Trans. Artif. Intell. **3**, 867–886 (2022)
13. Acheampong, F., Wenyu, C., Nunoo-Mensah, H.: Text-based emotion detection: advances, challenges, and opportunities. Eng. Rep. **2**, e12189 (2020)
14. Murthy, A., Kumar, K.: A review of different approaches for detecting emotion from text. In: IOP Conference Series: Materials Science and Engineering, vol. 1110, p. 012009 (2021)
15. Saxena, A., Khanna, A., Gupta, D.: Emotion recognition and detection methods: a comprehensive survey. J. Artif. Intell. Syst. **2**, 53–79 (2020)
16. Shinn, N., Labash, B., Gopinath, A.: Reflexion: an autonomous agent with dynamic memory and self-reflection. ArXiv Preprint ArXiv:2303.11366 (2023)
17. Salichs, M., Malfaz, M.: A new approach to modeling emotions and their use on a decision-making system for artificial agents. IEEE Trans. Affect. Comput. **3**, 56–68 (2011)
18. Egges, A., Kshirsagar, S., Magnenat-Thalmann, N.: Generic personality and emotion simulation for conversational agents. Comput. Anim. Virtual Worlds **15**, 1–13 (2004)
19. Webster, J., Watson, R. Analyzing the past to prepare for the future: writing a literature review. MIS Q. xiii–xxiii (2002)
20. Ismail, M., Ariffin, S.M.Z.S.Z.: Adapting to learners' emotions through animated pedagogical agent. In: 2014 3rd International Conference on User Science and Engineering (i-USEr), pp. 164–167 (2014)
21. Chatzara, K., Karagiannidis, C., Stamatis, D.: An intelligent emotional agent for students with attention deficit disorder. In: 2010 International Conference on Intelligent Networking and Collaborative Systems, pp. 252–258 (2010,11)
22. Gu, X., Wang, Z., Zheng, S., Wang, W.: Design and implementation of the emotional pedagogical agent. In: 2009 International Conference on Artificial Intelligence and Computational Intelligence, vol. 4, pp. 459–462, November 2009
23. Damjanovic, V., Kravcik, M., Devedzic, V.: eQ: an adaptive educational hypermedia-based BDI agent system for the semantic web. In: Fifth IEEE International Conference on Advanced Learning Technologies (ICALT'05), pp. 421–423, July 2005
24. De Silva, P.R., Madurapperuma, A.P., Marasinghe, A., Osano, M.: Integrating animated pedagogical agent as motivational supporter into interactive system. In: The 3rd Canadian Conference On Computer and Robot Vision (CRV'06), p. 34, June 2006

25. Sun, Y., Li, Z., Xie, J.: A formal model of emotional pedagogical agents in intelligent tutoring systems. In: 2013 8th International Conference On Computer Science & Education, pp. 319–323, April 2013
26. Rodrigues, R., Silva, R., Pereira, R., Martinho, C. Interactive empathic virtual coaches based on the social regulatory cycle. In: 2019 8th International Conference on Affective Computing and Intelligent Interaction (ACII), pp. 69–75 (2019)
27. Moga, H., Sandu, F., Danciu, G., Boboc, R., Constantinescu, I.: Extended control-value emotional agent based on fuzzy logic approach. In: 2013 11th RoEduNet International Conference, pp. 1–8, January 2013
28. Lone, M.,et al.: Self-learning chatbots using reinforcement learning. In: 2022 3rd International Conference on Intelligent Engineering and Management (ICIEM), pp. 802–808, April 2022
29. Katayama, S., Mathur, A., Broeck, M., Okoshi, T., Nakazawa, J., Kawsar, F.: Situation-aware emotion regulation of conversational agents with kinetic earables. In: 2019 8th International Conference on Affective Computing and Intelligent Interaction (ACII), pp. 725–731 (2019)
30. Chao, F., Lin, C., Kui, J., Zhonglin, W., Bing, Z.: Research on decision-making in emotional agent based on reinforcement learning. In: 2016 2nd IEEE International Conference on Computer and Communications (ICCC), pp. 1191–1194 (2016,10)
31. Jiwen, H., Jinfeng, L., Zhonglin, W., Yi, L. A framework for modeling emotion in synthetic agent. In: Proceedings of 2014 IEEE Chinese Guidance, Navigation and Control Conference, pp. 1151–1154, August 2014
32. Urban, G., Adamatti, D.: An adaptive agent approach using personality and emotions. In: 2018 XLIV Latin American Computer Conference (CLEI), pp. 215–223, October 2018
33. Hoffmann, C., Linden, P., Vidal, M.: Creating and capturing artificial emotions in autonomous robots and software agents. J. Web Eng. **20**, 993–1030 (2021)
34. Yusuf, R., Sharma, D., Tanev, I., Shimohara, K.: Individuality and user-specific approach in adaptive emotion recognition model. In: 2017 International Conference On Biometrics and Kansei Engineering (ICBAKE), pp. 1–6 (2017)
35. Suzuki, K., Camurri, A., Ferrentino, P., Hashimoto, S.: Intelligent agent system for human-robot interaction through artificial emotion. In: SMC'98 Conference Proceedings. 1998 IEEE International Conference On Systems, Man, And Cybernetics (Cat. No. 98CH36218), vol. 2, pp. 1055–1060, October 1998
36. Guerrero-Vásquez, L., Chasi-Pesantez, P., Castro-Serrano, R., Robles-Bykbaev, V., Bravo-Torres, J., López-Nores, M.: AVATAR: implementation of a human-computer interface based on an intelligent virtual agent. In: 2019 IEEE Colombian Conference on Communications and Computing (COLCOM), pp. 1–5, June 2019
37. Erol, B., Majumdar, A., Benavidez, P., Rad, P., Choo, K., Jamshidi, M.: Toward artificial emotional intelligence for cooperative social human-machine interaction. IEEE Trans. Comput. Soc. Syst. **7**, 234–246 (2020)
38. Patil, V., Hadawale, O., Pawar, V., Gijre, M.: Emotion linked AIoT based cognitive home automation system with sensovisual method. In: 2021 IEEE Pune Section International Conference (PuneCon), pp. 1–7, December 2021
39. Mowafey, S., Gardner, S.: A novel adaptive approach for home care ambient intelligent environments with an emotion-aware system. In: Proceedings of 2012 UKACC International Conference on Control, pp. 771–777 (2012)
40. Wang, D., Tan, A.: Mobile humanoid agent with mood awareness for elderly care. In: 2014 International Joint Conference on Neural Networks (IJCNN), pp. 1549–1556, July 2014

41. Dutta, E., Bothra, A., Chaspari, T., Ioerger, T., Mortazavi, B.: Reinforcement learning using EEG signals for therapeutic use of music in emotion management. In: 2020 42nd Annual International Conference of the IEEE Engineering In Medicine & Biology Society (EMBC), pp. 5553–5556 (2020,7)
42. Mowafey, S., Gardner, S.: Towards ambient intelligence in assisted living: the creation of an intelligent home care. In: 2013 Science and Information Conference, pp. 51–60, October 2013
43. Zhao, W., Zhao, Y., Lu, X., Wang, S., Tong, Y., Qin, B.: Is ChatGPT Equipped with Emotional Dialogue Capabilities? ArXiv Preprint ArXiv:2304.09582 (2023)

Author Index

Printed in the United States
by Baker & Taylor Publisher Services